04-CTS-12S



# Strategies for Small Wastewater Systems for Risk Reduction and Safeguarding Assets

by:

**Charles Herrick, Ph.D.**
Stratus Consulting

**Joy Barrett, Ph.D.**
Rural Community Assistance Partnership

*2009*

WERF

The Water Environment Research Foundation, a not-for-profit organization, funds and manages water quality research for its subscribers through a diverse public-private partnership between municipal utilities, corporations, academia, industry, and the federal government. WERF subscribers include municipal and regional water and wastewater utilities, industrial corporations, environmental engineering firms, and others that share a commitment to cost-effective water quality solutions. WERF is dedicated to advancing science and technology addressing water quality issues as they impact water resources, the atmosphere, the lands, and quality of life.

For more information, contact:
Water Environment Research Foundation
635 Slaters Lane, Suite 300
Alexandria, VA 22314-1177
Tel: (703) 684-2470
Fax: (703) 299-0742
www.werf.org
werf@werf.org

This report was co-published by the following organization. For non-subscriber sales information, contact:

IWA Publishing
Alliance House, 12 Caxton Street
London SW1H 0QS, United Kingdom
Tel: +44 (0) 20 7654 5500
Fax: +44 (0) 20 7654 5555
www.iwapublishing.com
publications@iwap.co.uk


Library of Congress Catalog Card Number: 2008940660
IWAP ISBN: 978-1-84339-352-8/1-84339-352-2



Stratus Consulting, Rural Community Assistance Partnership

The research on which this report is based was developed, in part, by the United States Environmental Protection Agency (EPA) through Cooperative Agreement No. R-830751-01 with the Water Environment Research Foundation (WERF). However, the views expressed in this document are not necessarily those of the EPA and EPA does not endorse any products or commercial services mentioned in this publication. This report is a publication of WERF, not EPA. Funds awarded under the Cooperative Agreement cited above were not used for editorial services, reproduction, printing, or distribution.

# ACKNOWLEDGMENTS

This research was funded by the Water Environment Research Foundation. The authors wish to thank the project subcommittee and Water Environment Research Foundation staff for their review and thoughtful input.

## Report Preparation

**Principal Investigators:**

Charles Herrick, Ph.D.
*Stratus Consulting*

Joy Barrett, Ph.D.
*Rural Community Assistance Partnership*

**Project Team:**

Charles Herrick, Ph.D.
*Stratus Consulting*

Joy Barrett, Ph.D.
*Rural Community Assistance Partnership*

Nicole Kalas,
*Stratus Consulting*

## Project Subcommittee

Julia Anastasio
*American Public Works Association*

Homer Emery
*San Antonio Water System*

Gail Hackney
*Arizona Rural Water Association*

Bill Kramer
*National Rural Water Association*

Mary Messec Smith
*Colorado Department of Public Health and Environment*

Gene Taylor
*U.S. Environmental Protection Agency*

## Water Environment Research Foundation Staff

**Director of Research**: Daniel M. Woltering, Ph.D.

**Program Director**: Claudio Ternieden

# ABSTRACT AND BENEFITS


**Abstract:**

Recent disaster events have spawned a new emphasis on domestic security and emergency preparedness. Governments at all levels are taking action to reduce their vulnerabilities and prepare for emergencies. A great deal has been written concerning security practices for large- and medium-sized water and wastewater systems. Some of these practices are relevant and applicable to small, rural, and tribal wastewater systems, but many are not.

This report identifies security-related practices applicable for small wastewater systems. The report focuses on security practices consistent with the technical, managerial, and financial capacity of small systems, and identifies a series of security-related "Practice Areas" that can be implemented in the near term with modest expenditure of financial and/or staff resources. The report also outlines a strategy to help small utilities define and implement programs for ongoing, sustainable security enhancement.


**Benefits:**

- Provides a resource for general audiences concerned with risk reduction and security in small municipalities, including system managers, board members, and community leaders.
- Enables small wastewater systems to implement immediate security enhancements consistent with their technical, financial, and managerial capabilities.
- Identifies security vulnerabilities typical of small wastewater systems.
- Provides a framework that small systems can use to plan and implement a long-term, sustainable program of security.
- Provides a framework to enable wastewater systems to work with communities, municipalities, and regional entities to cooperate for emergency response and security.
- Provides a framework through which wastewater utilities can coordinate security planning with other important management practices, such as asset management and strategic planning.
- Provides background on potential security-related funding sources for municipalities and wastewater utilities.
- Includes a hypothetical case study to illustrate how a wastewater utility can interact with neighboring localities and other municipal entities to plan a program of long-term security enhancement.


**Keywords:** Wastewater security, emergency response, small, rural, best practices.

# TABLE OF CONTENTS

# LIST OF TABLES

# Executive Summary

A great deal has been written concerning security practices for large- and medium-sized water and wastewater systems. Some of these practices are relevant and applicable to small, rural, and tribal wastewater systems, but many are not. Small systems tend to have characteristics which preclude them from adopting many of the practices employed by larger wastewater and water utilities.

For this project, the research team evaluated existing resources to identify security-related practices and approaches that meet the following conditions:

- **Pertinent to small, rural needs:** Enhancements and/or practices that address security-related issues prevalent among small, rural wastewater systems.
- **Feasible in small, rural context:** Security enhancements and/or practices that are consistent with perceived risk profiles and the technical, managerial, and financial capacity of small, rural, and tribal wastewater systems.
- **Applicable to the needs of the broader municipality:** Wastewater systems are commonly operated under the authority of a county, municipality, parish, special district, or tribal government. Security enhancements and/or practices should be framed to extend, enhance, or cooperate with other municipal and/or regional-scale needs and operations.

## Project Objectives and Focus

This report adopts a two-pronged approach with respect to risk reduction and security enhancement for small wastewater systems. First, the report identifies a series of security-related "Practice Areas" that can be implemented in the near term with modest expenditure of financial and/or staff resources. Second, the report outlines a strategy through which small utilities can plan and implement programs for long-term, sustainable security enhancement.

## Research Approach

To identify security-related practices consistent with the needs and capabilities of small systems, the research team 1) conducted in-depth interviews with 45 small, rural wastewater systems; 2) reviewed after-action assessment reports developed by the National Rural Water Association (NRWA) describing damage and response actions taken on behalf of systems impacted by Hurricanes Katrina and Rita (NRWA, 2005); 3) conducted a "listening session" with system operators from Florida, Mississippi, Louisiana, and Alabama who had experienced the 2005 Gulf Coast hurricane season; 4) conducted a series of regional workshops involving small system operators, public works directors, and other municipal leaders to assess security-related needs and practices, and to elicit feedback on approaches being considered by the project research team; and 5) conducted broad-based technical literature review addressing small, rural, and tribal wastewater systems, their distinctive characteristics, and typical security-related practices and vulnerabilities.

## Key Areas of Near-Term Security Enhancement

- Perimeter and appurtenance control: Many small, rural system operators express confidence that they have adequate fencing, alarms, locks, keys, and lighting systems to reduce vulnerabilities to a wide range of threats, including vandalism, terrorism, and "attractive nuisance" liability issues. Most system operators feel that they are able to do a good job of securing doors, gates, hatches, manholes, and other appurtenances that might enable access by malevolent individuals, including terrorists.
- Oversight of remote components and citizen watch programs: The widely dispersed nature of system components, low population densities, and relatively small law enforcement forces employed by many rural municipalities and tribal governments combine to make it difficult to patrol and maintain adequate surveillance over utility assets. Some small systems report that citizen watch and similar programs significantly enhance and extend their ability to detect and report suspicious activities or other events that could result in system malfunction.
- Hazardous material control and personnel training: Standard training and relatively simple procedures can be instituted to help secure chemicals and assure safe operations.
- Access to back-up power generators and training in generator operation procedures: Power loss is perhaps the most common ramification of many types of emergency situations. Back-up generators provide improved reliability of service and reduce the likelihood of sewage overflows.
- Lift station inspection, start-up, and by-pass procedures: Lift stations are essential to the operation of many wastewater collection systems, and are vulnerable to damage from a wide range of disaster situations. It is important that utility staff conduct regular inspection and preventive maintenance, which may help to reduce fouling and thereby optimize the system's emergency capacity. Similarly, system operators must be familiar with start-up and by-pass procedures in order to assess and repair lift stations, reduce down-time, and assure reliable operation.
- Mutual aid and assistance agreements: Many small systems operate under some type of mutual aid and assistance arrangement with other utilities and/or municipalities. Mutual aid and assistance provides a way for resource-limited organizations to obtain temporary access to equipment and technical capabilities needed to address disaster situations. In this context, many utilities interviewed for this project report close, effective, and long-term interaction with local volunteer and professional fire departments. Such relationships provide a strong foundation for cooperation during emergency events. Since drafting an original agreement can be expensive and time-consuming, many utilities chose to customize existing templates.
- Emergency Response Training, including training on system start-up and shut-down procedures: Many emergency situations will require systems to be fully or partially shut-down and/or restarted; it is essential that staff understands these procedures and do not create additional problems through improper shut-down or start-up.

## Coordination with the Broader Community

Utility security influences and is influenced by other municipal service and infrastructure providers. Utilities can combine security enhancement with other important managerial functions, such as strategic planning and asset management. When considered in cooperation with the wider municipality, such activities can facilitate overall risk reduction and long-term improvements to wastewater system security.

CHAPTER 1.0

# SMALL WASTEWATER SYSTEM SECURITY: BACKGROUND

## 1.1 Background

The Gulf Coast hurricanes of 2005 and horrific events of September 11, 2001 have spawned a new emphasis on domestic security and emergency preparedness. Governments at all levels are taking action to reduce their vulnerabilities and prepare for all types of emergencies, including unconventional disasters such as regional-scale weather events and terrorist attacks.

A great deal has been written concerning security practices for large- and medium-sized water and wastewater systems. Some of these practices are relevant and readily applicable to small, rural, and tribal wastewater systems, but many are not. Small, rural wastewater systems tend to have characteristics which preclude them from adopting or drawing upon many of the practices employed by larger wastewater and water utilities.

This report identifies security-related practices that are applicable for typical small, rural wastewater systems. The report adopts a two-pronged approach with respect to security enhancement for small wastewater systems. On the one hand, the report focuses on security practices that are consistent with the technical, managerial, and financial capacity of small systems. To this end, the report identifies a series of security-related "Practice Areas" that can be implemented in the near term with modest expenditure of financial and/or staff resources. On the other hand, the report adopts a forward-looking posture to help small utilities map strategies for long-term, ongoing, and sustainable security enhancement. This long-term strategy is based primarily on the identification of practices and investments that a utility can pursue in cooperation with other municipal and regional entities.

## 1.2 Small Wastewater Systems: A Profile

For many purposes, a water or wastewater system is considered small if it serves 10,000 or fewer people although systems serving fewer than 500 people are common and face some of the biggest challenges in providing adequate service. In the United States there are over 21,000 wastewater treatment systems. Approximately three-fourths (15,300) of these systems serve small communities, with 2,800 serving communities of between 3,500-10,000 people, 5,600 serving communities of between 1,000-3,500 people, and almost 7,000 serving communities with a population of less than 1,000. While there are numerous small wastewater systems in the United States, they serve a relatively small proportion (10%) of the sewered population. By way of comparison, about 500 large public wastewater systems provide service to about 62% of the sewered population. In other words, while large wastewater systems serve more people than small systems, small systems are far more prevalent than their large counterparts and comprise a large and critical aspect of the nation's overall community service and environmental protection infrastructure.

Many small wastewater treatment facilities and collection systems were constructed in the 1970s and 1980s, during a period in which federal funding was relatively abundant. The rated service life of a treatment plant is between 30-40 years; but other components within the system (pumps, computers, electrical equipment) have shorter service lives ranging from 10-25 years. This means that many small wastewater systems are approaching the end of their useful service lives and need expensive rehabilitation or replacement. Localities are struggling to pay the operating expenses of these aging facilities.

As the National Research Council (1997) reports, small communities can typically be divided into two groups: those in non-metropolitan areas and those in outlying suburbs or major metropolitan communities. Financial and other resources tend to be most limited in the non-metropolitan communities where average incomes are lower, unemployment is higher, and ratepayer bases are generally smaller. Small, rural communities tend to suffer from a lack of access to loan capital, a problem that impacts private systems more than publicly owned systems because they are not eligible to receive federal and state government grants. Economy of scale is also an issue: small tanks, pipes, and pumps can cost more per unit of wastewater processed than larger sizes and unit purchase quantities.

Small, rural wastewater systems tend to have fewer customers, fewer employees, and fewer miles of pipe than their large urban and suburban counterparts. However, smaller systems frequently provide service over a relatively large geographic area and operate and maintain an array of spatially dispersed components. In many cases, small wastewater systems are maintained by contract operators, who circulate staff and technicians among systems, but do not provide a full-time presence.

In general, small wastewater utilities have nowhere near the financial capacity of large private and municipal systems, and frequently lack the technical and managerial capacity of their large counterparts.

## 1.3 Research Orientation and Technical Approach

Over the past five years, water sector organizations have conducted extensive research and outreach efforts to identify, describe, and disseminate effective security-related practices for drinking water and wastewater utilities. While these efforts have resulted in some excellent and comprehensive guidance materials (see Table 1-1), they are not always applicable for small, resource-limited utilities and/or municipalities. From the perspective of small, rural, and tribal wastewater utilities, the existing portfolio of security-related guidance materials.

- Provides applicable approaches to vulnerability assessment, but relatively less "how to" information on security enhancement.
- Does not differentiate adequately between activities or initiatives that are suitable in the operational context of small systems, as opposed to large and medium systems.
- Tends to provide specialized, security-related material, rather than synthesized material appropriate for application by managers, operators, and community leaders.
- Provides a strong emphasis on terrorism, but relatively less focus on other, more likely issues.
- Provides a strong emphasis on emergency response, but relatively less focus on ongoing security enhancement.
- Does not provide adequate linkage with other state and federal mandates (e.g., environmental planning, asset management).

- Is focused predominantly on the needs of drinking water systems.

**Table 1-1. Examples of Existing Security-Related Guidance for Wastewater Systems.**

| |
|---|
| **American Lifelines Alliance, *Wastewater System Performance Assessment Guideline*:** Comprehensive resource; includes major infrastructure improvement as well as management practices. |
| **American Society of Civil Engineers, *Voluntary Security Guidance for Wastewater/Stormwater Utilities*:** Comprehensive resource, includes major infrastructure improvement and managerial practices. |
| **Association of Metropolitan Sewerage Agencies, *Asset Based Vulnerability Checklist for Wastewater Utilities*:** Booklet length tool to help managers identify and evaluate potential threats to their system, with short introductions to numerous potential emergency response and security enhancement actions. |
| **FlaWARN, *Emergency Response and Preparedness Best Practices for Florida's Water and Wastewater Systems*:** Extremely detailed reference source on security-related infrastructure, process, management systems, and emergency response procedures; strong emphasis on the establishment and utilization of mutual aid approaches. |
| **National Environmental Services Center for Small Communities, *Protecting Your Community's Assets: A Guide for Small Wastewater Systems*:** This booklet outlines a four-step process through which utilities (1) conduct an inventory of their critical assets; (2) identify the "most likely" threats to their systems; (3) identify vulnerable assets; and (4) prioritize corrective actions for identified vulnerabilities. |
| **National Rural Water Association, *Rural and Small Water and Wastewater System Emergency Response Plan Template*:** This booklet contains a series of templates, forms, and information summaries to help utilities prepare for emergencies. It does not address security enhancement per se, but includes a hazard ranking form and a critical components list. |
| **National Drinking Water Advisory Council, *Active and Effective Security Programs*:** A series of 14 organizational, operational, infrastructure-related, and external recommendations to improve water system security. |

For this project, it was necessary to evaluate existing resources to identify security-related practices and approaches that meet the following conditions:

- Pertinent to small, rural needs: Enhancements and/or practices that address security-related issues prevalent among small, rural wastewater systems.
- Feasible in small, rural context: Security enhancements and/or practices that are consistent with perceived risk profiles and the technical, managerial, and financial capacity of small, rural, and tribal wastewater systems.
- Applicable to the needs of the broader municipality: Wastewater systems are commonly operated under the authority of a county, municipality, parish, special district, or tribal government. Security enhancements and/or practices should be framed to extend, enhance, or cooperate with other municipal and/or regional-scale needs and operations.

To identify specific practices consistent with these criteria, the research team 1) conducted in-depth interviews with 45 small, rural wastewater systems; 2) reviewed after-action assessment reports developed by the National Rural Water Association (NRWA) describing damage and response actions taken on behalf of systems impacted by Hurricanes Katrina and Rita (NRWA, 2005); 3) conducted a "listening session" with system operators from Florida, Mississippi, Louisiana, and Alabama who had experienced the 2005 Gulf Coast hurricane season; 4) conducted a series of regional workshops involving small system operators, public works directors, and other municipal leaders to assess security-related needs and practices, and to elicit feedback on approaches being considered by the project research team; and 5) conducted broad-based technical literature review addressing small, rural, and tribal wastewater systems, their distinctive characteristics, and typical security-related practices and vulnerabilities.

Input from the utility survey and four utility workshops was critical in defining security-related needs, identifying security and risk reduction practices appropriate for small wastewater utilities, and helping to discern a project scope and reporting approach most applicable to the needs and circumstances of small utilities.

The first utility workshop was conducted as a "listening session" in Hattiesburg, Mississippi. At this event, participants were asked to share their experiences related to Hurricanes Katrina and Rita. Although participants were asked to comment on topics identified in the WERF RFP, the interaction process was not highly structured and participants were encouraged to speak freely regarding all matters of security-related concern. After review of session notes, inputs were organized in terms of 1) vulnerability assessments, 2) emergency response planning, 3) emergency communications, and 4) training. Mutual aid and assistance was also a salient topic and was discussed in terms of emergency response planning and training.

Input from the Hattiesburg session and the project literature review was used to develop a survey strategy focused on small system operators and/or management, municipal public works officials, and other relevant individuals. The questionnaire for this survey is included as Appendix A of this report. In addition to providing project background information, survey inputs were used to develop a provisional list of prevalent security-related issues, perceived security-related strengths, and areas in which utilities and municipalities seek guidance.

The second workshop was held in Buena Vista, Colorado. At this session participants were briefed on summary findings from the utility survey, including 1) prevalent emergency and security-related events, 2) areas of perceived and experienced security vulnerability, and 3) areas of perceived and experienced security readiness. Specifically, participants reacted to a list of security-related needs, a list of recognized security or risk reduction practices; and a conceptual discussion of how security practices should be framed and formatted to be of most use to small utilities. Participants also provided specific guidance on areas of effective security practice.

Participants at the Buena Vista workshop were especially focused on the need and opportunity to coordinate security planning with other utility management processes, such as insurance application reviews and financial accounting audits. Participants also emphasized that project outputs should be framed for political and municipal leaders, municipal planning staff, as well as utility staff. Participants were unanimous in the view that security-related guidance materials be kept short and simple, while providing easy access to resources of greater depth and technical specificity.

The third workshop was conducted in Sumter, South Carolina. At this session, participants were again requested to respond to lists of security issues and risk reduction practices. Based on conceptual input obtained through the Buena Vista workshop, the research team developed templates for a series of broadly effective security-related practices. During the workshop, session participants were allotted time to review each of the templates. Participants provided comments on 1) the relevance of each topic; 2) the level of detail, scope, and technical appropriateness of the content; and 3) the format of the material. Input from the Sumter workshop resulted in substantial shortening and simplification of several of the security practice area write-ups; addition of relevant detail to almost all of the write-ups; and elimination of several of the topic areas.

Revised security practice templates were presented at the fourth, and final, workshop, held in Eureka, California. Participants provided positive feedback on the basic content, level of detail, technical orientation, and presentational format of the best practice templates. Echoing participants at the Buena Vista workshop, participants in the Eureka session strongly endorsed the importance of providing a framework through which to integrate security with other utility management practices and regulatory mandates. Participants were also highly focused on the challenge of emergency communications, and articulated a "least common denominator" standard for security and risk reduction guidance wherein materials shall be written to be fully comprehensible to wastewater system oversight officials.

CHAPTER 2.0

# SECURITY NEEDS FOR SMALL WASTEWATER SYSTEMS

## 2.1 Introduction

This chapter summarizes small wastewater system security-related issues, security and risk reduction measures, and describes ten broadly effective security and risk reduction practices applicable for near-term utilization in the context of small and rural wastewater systems. Section 2.2 summarizes key inputs from the project survey and workshop series, including risk factors and risk reduction and emergency preparedness measures. Section 2.3 provides a brief description of security enhancement measures found to be common among small systems, followed by a series of ten widely effective security practices, formatted to help utility and municipal personnel identify and act upon opportunities for near-term risk reduction.

## 2.2 Security and Small Wastewater Systems: Status, Typical Characteristics, and Reasons to Adopt a Proactive Posture

In many ways, small wastewater systems are basically reactive when it comes to risk reduction, security enhancement, and emergency response planning. This is perhaps surprising since over half of the utilities interviewed for this project report that they have experienced some type of natural, technological, or human-caused emergency event within the past 10 years. Slightly over one-half of small systems have an emergency response plan (ERP), many of which were developed and are executed under the authority of a larger governmental entity, such as a municipality, county, or tribe. However, few systems (18%) have conducted training exercises or drills using their ERP; and less than half are aware of incident command system (ICS) roles and responsibilities with neighboring entities or governmental units. Approximately half claim to employ some type of mutual aid or assistance agreement with nearby utilities, a large proportion of which appear to be non-specific, verbal, handshake agreements.

A significant majority of systems rate their emergency preparedness as "good, but could be enhanced;" however, the basis for this assertion is unclear because fewer than one-quarter have ever activated their emergency plans. Slightly over one-third of the utilities interviewed for this project have completed vulnerability assessments; approximately 45% of which have been updated within the past two years. Based on our survey, it appears that less than 40% of small wastewater utilities have conducted asset inventories. Tables 2-1 and 2-2 outline key small system security-related activities and characteristics.

**Table 2-1. Summary of Small Wastewater System Emergency and Security Management Activities.**

| Have recent experience with a disaster event | Have a written emergency response plan | Have conducted security-related or emergency response drills | Have conducted a vulnerability assessment | Have a mutual assistance agreement in place | Have conducted an asset inventory |
|---|---|---|---|---|---|
| 64% | 59% | 18% | 39% | 47% | 46% |
| | | | Note: Few report using existing guidance materials or templates | Note: About half are informal, handshake agreements | |

**Table 2-2. Utility Self-Assessment of Security Status by Major Component and Management Areas.**

| Component or Activity Area | Percentage Deemed "Adequate" | Percentage Deemed "Inadequate" |
|---|---|---|
| Treatment | 77% | 23% |
| Collection System | 70% | 30% |
| Perimeter Security | 63% | 38% |
| Chemical and Hazardous Material Control | 87% | 13% |
| Emergency Supply Arrangements | 92% | 8% |
| Personnel | 78% | 22% |
| Computers and Data Systems | 86% | 14% |
| Records, Maps, and Information | 88% | 12% |

Until recently, small wastewater systems have not been subject to governmental and/or regulatory mandates to conduct security assessments or emergency response planning activities. While there remain no regulatory requirements for small wastewater systems to conduct security-related activities, the U.S. Department of Agriculture, Rural Utilities Service (RUS) requires development grant recipients to certify that they have conducted vulnerability assessments and developed emergency response plans.

Beyond governmental prescriptions, there are good reasons for small systems to adopt a more proactive approach to security planning. In the first place, wastewater system vulnerability to extreme events is a well-documented phenomenon. Over the past two decades, 100-year floods, earthquakes, and other natural disasters have heavily damaged wastewater treatment plants and collection systems. Severe wind events, ice storms, and accidents regularly result in electrical power outages, especially for smaller systems operating without on-site staff or large maintenance crews. As illustrated by Table 2-3, the utilities interviewed for this project have experience with a variety of significant emergency events.

**Table 2-3. Disaster Events Experienced by Utilities in this Study.**

| | |
|---|---|
| Rainfall overflow | Ice storm resulting in power outage |
| Chemical dumps | Hurricane resulting in prolonged (11 day) power outage |
| Earthquake resulting in collection line rupture | High wind event, resulting in 16 hour power outage |
| Gasoline spill | Severe storm, resulting in system overflow and power outage |

While surveys suggest that many Americans believe that future terrorist attacks are likely, few believe their community will be a target. (WDHFS, 2004) This is especially true of citizens living in small and rural municipalities. Likewise, small system operators tend not to view terrorism as an especially salient threat. Perceived risk profiles notwithstanding, litigation over terrorist-induced harm should be a matter of concern for wastewater utilities and their host municipalities. Court rulings have found that utilities must exercise reasonable care in operating and maintaining their systems. (Anderson and Schultz, 2005) The definition of "reasonable care" is key in determining liability. As more utilities implement vulnerability assessments and security-related improvements, it may be argued that the definition of reasonable care is evolving to include practices and/or installation of security measures that only a short time ago were rarely found in wastewater systems.

Survey findings, workshop inputs, and other research from this study combine to suggest a core set of security-related needs common to many small, rural wastewater systems.

- Electrical power loss: Based on our survey, most small systems have experienced prolonged (at least 1 day) power outages, resulting in shut-down of key system components and interruption of services. In several cases, these power outages lasted for over a week.
- Emergency fuel: Regional-scale disasters can result in widespread fuel shortages, severely limiting back-up power capabilities, operation of service and repair vehicles, and other essential functions.
- Damage to lift stations: Lift stations can be damaged by a wide variety of emergency events, including hurricanes, sea water infusion, floods, earthquakes, ice storms, mud slides, and malevolent human interference.
- Resilient, self-contained emergency communications systems: Wastewater utilities that managed their way through the 2005 Gulf Coast hurricane season express frustration with extended failures in standard modes of communication, including cellular phones.
- Written emergency response plans: Many small, rural wastewater systems do not have written emergency plans or emergency procedures for specified components.
- Confusion regarding ICS roles and responsibilities: Many small systems seem unaware and/or confused about ICS roles and responsibilities, including emergency authorities within their host municipality or district.
- Access to security-related funding: Many small wastewater systems operate with little operational reserve for addressing security-related and other non-routine matters.
- Emergency communications: During an emergency it is critical to direct employees and inform the media, customers, and the public. Effective response actions must be communicated quickly and clearly because inappropriate reactions can worsen the impact of the initial event.

## 2.3 Security Practices Applicable to the Needs of Small Wastewater Systems

The resource-limited nature of many small systems constrains the type of security-related activities they can realistically pursue. It is essential that security practices for small systems be scaled to address operational realities. The security-related practices outlined below are synthesized from our literature review and survey of small system operators, and are believed to be feasible given typical operational circumstances. While these practices do not constitute an overall security program, each will help to enhance a utility's security and emergency response capabilities. Security-related practices include the following:

- Enhanced operations and maintenance: A focus on routine maintenance activities can pay substantial dividends in terms of a system's resilience to many kinds of emergency situations. If components are in good operating condition, they will better withstand above-normal loadings and other stresses associated with emergency situations.
- Perimeter and appurtenance control: Many small, rural system operators express confidence that they have adequate fencing, alarms, locks, keys, and lighting systems to reduce vulnerabilities to a wide range of threats, including vandalism, terrorism, and "attractive nuisance" liability issues. Most system operators feel that they are able to do a good job of securing doors, gates, hatches, manholes, and other appurtenances that might enable access by malevolent individuals, including terrorists.
- Oversight of remote components and utilization of citizen watch programs: The widely dispersed nature of system components, low population densities, and relatively small law enforcement forces employed by many rural municipalities and tribal governments combine to make it difficult to maintain adequate surveillance over utility assets. Some small systems report that citizen watch and similar programs significantly enhance and extend their ability to detect and report suspicious activities or other events that could result in system malfunction.
- Hazardous material control and personnel training: Standard training and relatively simple procedures can be instituted to help secure chemicals and assure safe operations.
- Access to back-up power generators and training in generator operation procedures: Power loss is perhaps the most common ramification of many types of emergency situations. Back-up generators provide improved reliability of service and reduce the likelihood of sewage overflows.
- Lift station inspection, start-up, and by-pass procedures: Lift stations are essential to the operation of many wastewater collection systems and are vulnerable to damage from a wide range of disaster situations. It is important that utility staff conduct regular inspection and preventive maintenance, which may help to reduce fouling and thereby optimize the system's emergency capacity. Similarly, system operators must be familiar with start-up and by-pass procedures in order to assess and repair lift stations, reduce down-time, and assure reliable operation.
- Mutual aid and assistance agreements: Many small systems operate under some type of mutual assistance arrangement with other utilities and/or municipalities. Mutual assistance provides a way for resource-limited organizations to obtain temporary access to equipment and technical capabilities needed to address disaster situations. In this context, many utilities interviewed for this project report close, effective, and long-term interaction with local volunteer and professional fire departments. Such relationships provide a strong foundation for cooperation during emergency events.
- Emergency Response Training, including training on system start-up and shut-down procedures: Many emergency situations will require systems to be fully or partially shut-down and/or restarted; it is essential that staff understand these procedures and do not create additional problems through improper shut-down or start-up.
- Emergency Communications: It is advised that wastewater utility emergency response plans contain a section on emergency communications procedures and protocols. Inappropriate and/or ineffective messages and communications approaches can actually worsen the impact of a disaster or emergency event. Communications plans should include 1) internal and

partner contact lists and protocols, 2) public outreach and information dissemination, and 3) authorizations and protocols for dealing with the media.

The summary templates that follow provide "how to" information on the security "practices." Formatted for stand-alone use, each "Security Practice Area" template begins with a short description of the activity and an explanation of its importance to the utility and/or municipality. Each template outlines recommended activities, and as necessary, provides a URL or link to relevant guidance from leading technical institutions. The templates also identify opportunities for a wastewater utility to coordinate with other local, municipal, or regional organizations that may have complementary needs. Finally, each template includes citations for additional information.

**Security Practice Area 1: Operations and Maintenance.**

| | |
|---|---|
| **Description** | A focus on routine maintenance activities can pay substantial dividends in terms of a system's resilience to many kinds of emergency situations. If components are in good operating condition, they will better withstand above-normal loadings and other stresses associated with emergency situations. |
| **Procedures and Practices** | Maintenance needs are unique to individual wastewater systems. Apart from placing a general emphasis on good maintenance practices, it makes sense for a utility to conduct a formal asset inventory, including information on the condition, age, service history, and useful life of system components. Such an inventory will help in planning maintenance schedules and expenditures, but will also help your utility and/or municipality to comply with GASB #34 and related requirements. For more on this see http://www.epa.gov/safewater/smallsystems/pdfs/guide_smallsystems_asset_mgmnt.pdf, especially pages 5-17. Selected O&M activities that will contribute to system security and risk reduction include the following:<br><br>♦ Develop an understanding of the time of travel of the wastewater through the treatment facility at different flows to know how long it would take for contaminants to move from process to process and through the treatment plant. (ASCE, 2004)<br>♦ Inspect processes and overall operations daily, being vigilant to recognize any treatment system upset that may be related to contamination of the sewage or chemicals that are added to the treatment process. (ASCE, 2004)<br>♦ As applicable, rotate standby tanks, basins, mechanical equipment and electrical components into an operating mode routinely. Rotation allows maintenance activities to be conducted routinely to ensure that standby capacity is readily available when needed. (ASCE, 2004)<br>♦ Store critical spare equipment and parts in a secure location away from the operating equipment. This protects the equipment from a malevolent act that is directed at damaging the operational systems and allows the utility to quickly restore operations after an event occurs. (ASCE, 2004)<br>♦ Maintain complete records of system operating and maintenance data.<br>♦ Obtain and assure easy access to all "as built plans" from engineers. |
| **Coordination and Linkage** | Systems should establish and maintain relationships with contractors and vendors to assure prompt priority service, especially during emergency events. |
| **For More** | ASCE / AWWA / WEF, 2004. Interim Voluntary Guidance For Wastewater / Stormwater Utilities http://www.asce.org/files/pdf/wisc/2.pdf<br><br>U.S. Environmental Protection Agency, 2008. Check Up Program for Small Systems (CUPSS) http://www.epa.gov/safewater/cupss/casestudies.html |

## Security Practice Area 2: Perimeter and Appurtenance Control.

**Description**

Purposeful acts may include a direct attack on infrastructure such as a treatment plant, pump station, interceptor, or collection system. Additionally, a sewer system and its appurtenances (e.g., manholes, stormwater inlets) may provide a vandal or terrorist with a means to attack another target through the use of an explosive or flammable substance or by physical human entry into the sewer system. Purposeful and unintentional disposal of substances such as oils, pesticides, and herbicides into residential and non-residential sink drains, water closets, and floor drains has caused treatment plant upsets and threatened the health of utility workers. More dramatic, however, has been the dumping of chemical wastes from industrial sources into sewage collection systems, resulting in loss of treatment capabilities and destruction of infrastructure and surrounding properties.

**Procedures and Practices**

Utilities should adopt obvious and cost-effective physical improvements, such as perimeter fencing, tamper-proof manhole covers, and locking valve boxes.

- Invest in fencing, gates, illumination, and clear zones around remote and critical facilities. For more on this see pages A-1 through A-10 of ASCE/AWWA/WEF, 2006; and pages 4-4 through 4-17 of ASCE/AWWA/WEF, 2004.
- Tack welding provides a low-cost, near-term security measure, useful in areas of the collection system known for illegal dumping into manholes and areas prone to theft of manhole covers.
- Bolt-type locking devices consist of bolts that anchor the manhole cover to the frame. These devices prevent the cover from being removed, as well as preventing access into the manhole. Pan locks prevent access into the manhole as well as preventing dumping into the collection system. Securing manhole covers also prevents vandals from removing covers causing hazardous conditions for motorists and pedestrians and discourages thefts of manhole covers by collectors.
- Secure valve and gate operators in their operating configuration to avoid malicious tampering or an unintentional action by an employee. Hardened chains and locks can be used.
- Keep electrical and control boxes locked at all times to prevent unauthorized access.

**Coordination and Linkage**

If applicable, involve citizen watch groups and other municipal staff in an observational capacity.

Provide access instructions and keys to emergency responders, including volunteer fire personnel.

Develop close relationships with local law enforcement agencies; make sure they know where critical assets are located. Request that they add your facilities to their routine rounds.

**For More**

EPA Water and Wastewater Security Product Guide
http://cfpub.epa.gov/safewater/watersecurity/guide/

ASCE/AWWA/WEF, 2006. Guidelines for the Physical Security of Wastewater/Stormwater Utilities.
http://www.asce.org/static/1/redirect.cfm?prmType=WISE&prmFile=20062/00_Complete_Document

ASCE / AWWA / WEF, 2004. Interim Voluntary Guidance For Wastewater/Stormwater Utilities
http://www.asce.org/files/pdf/wisc/2.pdf

## Security Practice Area 3: Auxiliary Power.

**Description**

A wide variety of emergency and disaster events can result in power outage. In preparation for an emergency, it is important to know generator capacity load, fuel consumption, and to be familiar with start-up procedures. Generators should be load-tested and serviced frequently. Another important factor to consider is how to secure fuel after the disaster event.

**Procedures and Practices**

Best practices for back-up power include:

- Estimate generator capacity load and fuel consumption of your auxiliary power system.
- Capacity load: estimate the capacity load of your back-up generator; for more on this see: http://watersc.org/pdf/FlaWARNBestPracticesJuly2006.pdf, especially pages 110-131. Affix a label prominently to the generator displaying the capacity load.
- Fuel consumption: estimate the hours of service a generator will provide when the tank is topped off; for more on this see: http://watersc.org/pdf/FlaWARNBestPracticesJuly2006.pdf, especially pages 35 and 128-129.
- Develop a power failure plan that identifies staff responsibilities including systematic shut-down of non-critical equipment and systems that could reduce loads on the generator.
- Establish mutual aid and assistance relationships that include prompt access to back-up power (see Security Practice Area – 7, this document).
- Generators should be available and tested under load each month to verify operation. (NESCSC, 2002).
- Generators should be serviced at least annually by a certified professional.
- Make sure that the service technician checks belts, batteries, and coolant, and has assured that the generator can accommodate anticipated loads and run for several days with no problems.
- Generator accessories such as connecting cords, extra wire, receptacles and plugs should be checked for availability. Support equipment should be packaged and attached to the generator. Assure that equipment that may need to be powered by auxiliary units can accommodate cable hook ups, rather than hard-wired connections.
- Every 3-5 weeks main breakers and automatic transfer switches need to be tested for proper function.
- If any problems are found with the generator, have it fixed immediately.
- Develop plan to secure fuel supplies during an emergency; contact several fuel vendors with estimated fuel volumes required if a storm occurs.
- Consider the vendors' physical location and potential delivery restrictions in emergency event.
- Ideally, power supplies should be located in places where they are least likely to be affected by disaster events. (NESCSC, 2002)
- Develop or adopt procedures and conduct training sessions on generator start-up and troubleshooting; for more see http://www.flawarn.org/resources/uploads/public/Documents/BMPs.pdf especially Chapter 15.

**Coordination and Linkage**

County and municipal street and highway maintenance departments (and their contractors) own and operate relatively large, mobile generator units. Make sure that local and regional electric utilities are aware of the standby and/or emergency power requirements of the wastewater system, so that they can assure and safeguard adequate provision of emergency power.

**For More**

FlaWARN, 2006. FlaWARN Emergency Response and Preparedness Best Management Practices for Water and Wastewater Systems http://www.flawarn.org/resources/uploads/public/Documents/BMPs.pdf.

## Security Practice Area 4: Lift Station Resilience.

**Description**

Disabled lift stations can create many problems in the wastewater system. Flooding of pump and lift stations can have a severe negative effect on the operation of the collection system. Lift station and pump station failures are also caused by power failures, frequently caused by downed trees and limbs coming into contact with overhead power lines. Aside from an adequate workforce, three important tools for lift station recovery include portable and fixed generators to keep the waste stream flowing toward the treatment plant (see Practice Area #3), pumps for bypassing damaged systems, and vacuum and jetting trucks to clear blockages. Mutual aid and assistance agreements and/or advance agreements with outside contractors can provide access to such equipment.

**Procedures and Practices**

Effective practices for lift station protection include:

- Develop post-event damage assessment checklists for individual lift stations; train staff in their utilization. For more see: http://watersc.org/pdf/FlaWARNBestPracticesJuly2006.pdf, especially pages 37, 71, and 135-142.
- Secure access points and control panels at lift stations with tamper resistant locks. Protect structures from collision damage by means of concrete bollards or jersey barriers. Install alarms and test them regularly. (NESCSC, 2002)
- Mitigate or avoid possible flooding issues through installation of risers to increase the top grade of lift station structures; or through installation of temporary or semi-permanent barriers around vulnerable lift stations and pump stations. Protective barriers can be built from sand bags and other materials to divert floodwaters away from the station structure. Stage barrier construction materials (sand bags) near vulnerable sites. For more on this see: http://watersc.org/pdf/FlaWARNBestPracticesJuly2006.pdf, especially pages 39-41.
- It is prudent to conduct frequent power line right-of-way inspection and maintenance to eliminate overgrowth near power supplies. While outside utility control, this may be accomplished through cooperative agreements or under the aegis of regional emergency preparedness plans.
- Consider the purchase and installation of submergence-rated equipment when performing routine system maintenance or replacement, or planning for system expansion and/or upgrades.

**Coordination and Linkage**

Work with electrical power transmission and/or distribution authorities to assure sufficient vegetation control and maintenance of power line right-of-ways.

Damages to power lines feeding lift stations can be significantly reduced through the use of underground power service. When planning new developments and/or upgrades of existing systems, utilities should consider underground power. Coordination opportunities may occur during roadway construction, drinking water system expansion, or other municipal ventures.

**For More**

FlaWARN, 2006. Emergency Response and Preparedness: Best Management Practices for Water and Wastewater Systems
http://www.flawarn.org/resources/uploads/public/Documents/BMPs.pdf.

American Lifelines Alliance, 2004. Wastewater System Performance Assessment Guideline
http://www.americanlifelinesalliance.org/pdf/aaaALAWWGuidelines070104DraftFinal.pdf.

**Security Practice Area 5: Neighborhood Watch and Security Enhancement of Remote Assets.**

| | |
|---|---|
| **Description** | Neighborhood Watch is a proven, nationwide program that reduces crime through the voluntary efforts of observant citizens. The concept can be applied to rural wastewater utilities. Many small, rural systems have unstaffed, remote assets or facilities. Organized neighbors can assist utilities in protecting these assets from vandals, criminals, and terrorists. They can also be trained to watch for storm damage or other disaster-related concerns, thereby helping the utility to locate potential problems and to react quickly in the case of damage. By devising effective neighborhood watch policies and outreach materials, wastewater utilities can enhance their asset and property protection, while strengthening customer relations. |
| **Procedures and Practices** | Best practices for citizen involvement include:<br>♦ Make sure citizen watch volunteers know who to call in case of an emergency or suspicious activity; develop and distribute emergency contact lists.<br>♦ Develop and circulate a simple list of "things to watch out for."<br>♦ Regularly send out information with bills to explain ways that customers can help protect the community's investment.<br>♦ Work with local law enforcement agencies to establish a neighborhood watch program that specifically includes wastewater system facilities and components.<br>♦ Make sure that utility staff and contractors wear badges or uniforms and drive clearly marked vehicles to help citizen watch participants discern who belongs on utility premises and who does not. |
| **Coordination and Linkage** | Municipal police, county sheriff, volunteer fire departments, county road crews, and telephone linemen are all "out and about" and can be trained to recognize suspicious or abnormal conditions. |
| **For More** | EPA, 2003. Water Watchers: Helping to Protect Your Local Water System<br>http://www.epa.gov/region01/eco/drinkwater/pdfs/waterwatchers_revised2.pdf<br>National Crime Prevention Council, 2002. Starting a Neighborhood Watch<br>http://www.ncpc.org/cms/cms-upload/ncps/files/nwstart.pdf. |

## Security Practice Area 6: Emergency Response Plan (ERP).

**Description**

The purpose of an ERP is to provide a utility with standardized response and recovery protocols to prevent, minimize, and mitigate damage resulting from emergencies or disasters of human-caused or natural origin. The ERP should outline the communication and coordination that would occur between the utility and local emergency response personnel, including police, fire, and public health officials. The plan should be developed in consultation with local emergency management personnel and first responders; this communication provides the utility with knowledge of local resources and begins the agency coordination necessary in responding to an emergency. The ERP should also identify equipment and resources to assist the utility in responding to and recovering from an emergency event.

The ERP, as well as other materials that may be needed during an emergency, must be kept up-to-date in a hardcopy format and available to the utility's staff. Electronic formats of ERPs, maps, manuals, and other documents may not be accessible during extended power outages. Key procedures (e.g., generator start-up, lift station shut-down) should be printed on waterproof, plastic stock to facilitate field use during adverse conditions. If applicable, the wastewater ERP should become an "annex" or appendix to a citywide ERP.

**Procedures and Practices**

A wastewater utility ERP may be organized into the nine core elements outlined below:

- Introduction. This section describes the purpose, goals, regulatory requirements and overall organization of the ERP.
- General Emergency Planning Information: This section describes the utility's emergency planning partnerships; mutual aid and assistance agreements; and emergency response policies, procedures and documents.
- System Specific Information: This section contains an overview of the wastewater system including a description of the collection system, pump stations, and a discussion of the treatment facilities and processes. System maps should be included to aid in understanding system capacities and inter-relationships of system components. Emergency resources, such as backup power supplies and redundant facilities, should also be listed in this section.
- Roles and Responsibilities: This section should describe the jurisdiction's incident command system (ICS); and the wastewater system structure for incident command.
- Communications Procedures: (See Practice Area #9).
- Personnel Safety: This section should address 1) evacuation and shelter planning, 2) off-site protective actions, and 3) first aid and emergency medical treatment.
- Emergency Storage and Disposal Provisions for Contaminated Wastewater and Biosolids: This section should identify storage and treatment options available during a contamination episode. Templates for effective display of such information are available through WERF, 2004, page 3-21.
- Equipment and Supplies: This section should include 1) an emergency equipment list, 2) VHF radio communications instructions, 3) telephone equipment, 4) personnel protective and other emergency equipment, and 5) stop-gap communications equipment.
- Appendices can include various documents relevant to emergency response of the utility, such as the following: 1) maps and other relevant system information 2) mutual aid and assistance agreements 3) example public notices and press releases 4) damage assessment checklists 5) threat documentation forms 6) contamination threat evaluation worksheets 7) generic site characterization plans 8) site characterization report forms 9) equipment lists 10) interconnects with other wastewater systems (for more see WERF, 2004, page 3-21).

**Coordination and Linkage**

Utilities should work with jurisdictional partners (e.g., mayors, public utility directors) to determine ICS expectations with respect to their roles in county, township, or municipal ERPs. If applicable, wastewater utilities should review municipal procedures or protocols that should (and do not) include utility involvement. Make sure the county, local fire department, and other applicable organizations each have a copy of the wastewater ERP. Wastewater systems may be required to assign staff to the City or County Emergency Operations Center to coordinate with drinking water, public health, or other departments.

**Security Practice Area 6: Emergency Response Plan (ERP).**

| | |
|---|---|
| **For More** | Water Environment Research Foundation, 2004. Emergency Response Plan Guidance for Wastewater Utilities<br><br>National Rural Water Association, 2002. Rural & Small Water and Wastewater System Emergency Response Plan Template.<br><br>EPA Response Protocol Toolbox<br>http://www.epa.gov/safewater/watersecurity<br><br>AMSA Vulnerability Self-Assessment Tool (VSATTM): ERP Module that can be used to guide a wastewater utility through ERP development.<br><br>ASIS, 2003. International Disaster Preparation Guide.<br>www.asisonline.org/newsroom/crisisResponse/disaster.pdf |

### Security Practice Area 7: Mutual Aid and Assistance Agreements.

**Description**

By reaching out to neighboring utilities and other organizations, a wastewater utility can gain use of equipment, technical resources, and trained personnel that would not otherwise be available. Mutual aid and assistance agreements provide a formal means of documenting coordination efforts between utilities or other agencies and can greatly facilitate the exchange of resources during an emergency. They are based on the concept that resources, in most circumstances, are voluntarily provided, that there will be a reciprocal exchange if and when required, and that providing resources will not result in a profit to the providing party. Selected services include the following:

- Allowing solids processing at each other's facilities.
- Allowing treatment of wastewater at each other's facility if wastewater is brought in by tanker truck.
- Acquiring supplies and materials from another utility.
- Borrowing light and heavy equipment between utilities.
- Using staff from another utility.
- Utilities should explore the possibility of joining state-level Water and Wastewater Agency Response Networks (WARN); for more on this see www.nationalwarn.org.

Templates for mutual aid and assistance agreements can be downloaded from these web sites: FlaWARN: http://www.flawarn.org/resources/uploads/public/Documents/BMPs.pdf, at pages 164-168.

**Procedures and Practices**

Effective agreements tend to adhere to the following practices (from AwwaRF, 2006):

- Participation must be voluntary; agreements cannot create a "duty" to respond. Although response is not mandatory, assistance requests should take priority over routine matters.
- It must be clear that potential lenders cannot be held liable for failure to provide assistance.
- Resources can be withdrawn at any time, even if they have already been lent.
- Use, repair, and/or replacement costs are borne by the borrower.
- Employees must be willing to participate.
- Incident command is the responsibility of the borrower.
- Borrowers must be willing to comply with the lenders statutory requirements (if any).
- Inventory information needs to be shared, participants should develop a list of priority materials and personnel, and the lists should be updated frequently and shared among all participants.
- Parties to an agreement should develop a hierarchy of needs to help guide lenders and incident commanders, including step-by-step procedures on how to request aid.
- Mutual aid and assistance agreements (and supporting inventories) should be reviewed and updated at least annually.
- Costs and reimbursement are frequently a point of contention in the arena of mutual aid and assistance, with conflicts fairly common. It is important that participants (a) spell out detailed cost accounting and reimbursement arrangements, or (b) rely on trust and existing handshake arrangements, accepting the risk of financial dispute.

**Coordination and Linkage**

Partners in mutual aid and assistance agreements include law enforcement, fire rescue, emergency management agencies; local health and public works departments; water, solid waste, electric, telecommunications, and gas utilities; and chemical suppliers.

Mutual aid and assistance partners should hold periodic meetings to share information, renew contacts, update information, and review emergency response protocols.

**For More**

EPA Security Information Collaboratives Guide
http://www.epa.gov/nhsrc/pubs/625r05002.pdf

Nationalwarn.org: sample agreement
http://www.awwa.org/files/WARN/WARN%20Model%20Agreement%20Jul%202008.doc#sample.

American Water Works Association Research Foundation, 2006. Emergency Response and Recovery Planning for Water Systems: A Kit of Tools

**Security Practice Area 8: Process Chemical and Hazardous Material Management.**

**Description**

Wastewater treatment facilities use a variety of chemicals as part of the treatment process. Security risks associated with chemicals or other hazardous materials include theft and misuse, use of the wrong chemical or a contaminated chemical, facility or component breach or rupture and subsequent environmental or worker exposure, and damage to the chemical feed system. The following is a partial list of chemicals and other hazardous substances commonly found at wastewater treatment facilities and, possibly, remote facilities such as pump stations:

- Acetylene
- Activated carbon
- Alum
- Ammonia
- Calcium hypochlorite
- Caustic chlorine
- Chlorine dioxide
- Defoamers
- Diesel fuel
- Digester gas
- Ferric chloride
- Ferric sulfate
- Fuel oil
- Gasoline
- Hydrochloric acid
- Hydrogen peroxide
- Lime
- Liquid propane gas
- Lubricating oils
- Methanol
- Odor-masking agents
- Oxygen
- Ozone
- Paints and thinners
- Pesticides
- Polymers
- Sodium bisulfate
- Sodium hypochlorite
- Solvents
- Sulfur dioxide
- Sulfuric acid
- Welding gasses

**Procedures and Practices**

The following practices can help reduce risks associated with chemical transport, handling, and storage.

- Proper handling of chemicals
- Cleanup of spills
- Personal protective clothing and equipment
- Training
- Work with suppliers to establish a delivery protocol. For guidance see ASCE, 2006, pages 3-7 through 3-8 (link below).
- Contractually obligate suppliers to implement a "seal lock" at the point of origin, and reject deliveries that may have been tampered with.
- Follow up with manufacturer to confirm late or not delivered shipments.
- Inspect chemical storage areas, storage tanks, and feed equipment weekly.

**Coordination and Linkage**

Coordinate with local HazMat response team and fire department to assure they are aware of the types and quantities of chemicals used and stored at utility facilities.

Coordinate with local and regional medical providers to assure they are aware of the types and quantities of chemicals used and stored at utility facilities.

Coordinate with local and regional law enforcement to assure they are aware of the types and quantities of chemicals used and stored at utility facilities.

Develop a map of chemical storage locations to include in the utility and/or municipal ERP.

**For More**

EPA Alert on Chemical Accident Prevention and Site Security
http://www.epa.gov/ceppo/pubs/secale.pdf

ASCE / AWWA / WEF, 2004. Interim Voluntary Guidance For Wastewater / Stormwater Utilities
http://www.asce.org/files/pdf/wisc/2.pdf

**Security Practice Area 9: Emergency Communications.**

**Description**

It is advised that wastewater utility emergency response plans contain a section on emergency communications procedures and protocols. Inappropriate and/or ineffective messages and communications approaches can actually worsen the impact of a disaster or emergency event. Communications plans should include 1) internal and partner contact lists and protocols, 2) public outreach and information dissemination, and 3) authorizations and protocols for dealing with the media.

**Procedures and Practices**

The following procedures can be used to enhance emergency and risk-related communications efforts:

- Make a list of actions that may be required of elected or appointed officials during an emergency situation. As necessary, train oversight officials on critical aspects of emergency operations.
- Prepare materials to inform the public concerning the onset of an emergency or other abnormal situation, including appropriate actions to take and instructions on what not to do. To facilitate this, think of plausible emergency events and their likely outcome on utility operations, then think of questions that the public may have regarding what to do and how to do it.
- Define in advance who will act as the utility's media spokesperson, and who will serve as backup media spokesperson.
- Utilities should maintain their investment in traditional two-way radios (Moyer, 2007) and anticipate the need to utilize "primitive" communications methods such as "runners," visual signs and signals, and pre-designated bulletin boards.
- Mutual aid and assistance agreements can be a useful backup to assure adequate emergency communications. Keep in mind that neighboring utilities may be subject to the same emergency incident, and thus suffer the same types of failures as your system.

**Coordination and Linkage**

Emergency contact lists should contain primary and alternate contact information for, at a minimum, utility personnel, local, state and federal emergency management staff, key chemical and equipment suppliers, contractors and consultants, mutual aid partners, media contacts and sensitive customers.

**For More**

NLC, 2002. *Homeland Security: Practical Tools for Local Governments*, see pages 18-23.

WERF, 2008. Emergency Communications with Your Local Government and Community, see pages 5-1 through 5-22.

**Security Practice Area 10: Emergency Response Training.**

**Description**

Many state and local resources are available to conduct, and sometimes fund, training for utility staff. When considering training, it is a good idea to check with local police and fire departments, local emergency planning agencies, local health departments, and the Red Cross to inquire about available training. Also, nearby utilities may want to share in training sessions and contribute funding. It is important to train staff in skills outside of their normal duties to assure adequate operational capabilities during times of emergency. Important training topics include the following:

- Security awareness and understanding standard operating procedures
- Understanding the Emergency Response Plan
- Incident Command System (ICS)
- General emergency management training
- Introduction to utility systems and processes for emergency responders
- Equipment use
- 800 megahertz (MHz) Radios
- Generator operations
- HAZWOPER (Hazardous Waste Operations and Emergency Response) training
- Cross-training

**Procedures and Practices**

There are several types of training, some of which are feasible for small systems:

- Orientation and education sessions: These are regularly scheduled discussion sessions to provide information, answer questions, and identify security-related needs and concerns.
- "Tailgate" exercises: Similar to tabletop exercises, utility staff gather and discuss their responsibilities and how they would react to specified emergency scenarios. This is a cost-effective way to identify areas of overlap and confusion before undertaking more demanding training activities.
- Walk-through drills: The emergency management group and/or response teams actually perform their emergency response function (e.g., chlorine leak). This usually involves more people and is more thorough than a tabletop exercise.
- Functional drills: These drills test specific functions such as emergency notification or contaminant monitoring procedures, although not necessarily at the same time. Staff are asked to evaluate the system and identify problems.
- Full-scale exercises: A real-life emergency situation is stimulated as closely as possible. This exercise involves local emergency response personnel, management, and community response organizations. Such drills involve a great deal of planning, advance coordination, and can be relatively costly.

**Coordination and Linkage**

At a minimum, emergency response plan activation training should include fire departments (including volunteer fire fighters); other municipal, county, or parish departments; and organizations with which the utility or municipality has a mutual aid and assistance agreement.

**For More**

ASCE / AWWA / WEF, 2004. Interim Voluntary Guidance For Wastewater / Stormwater Utilities http://www.watersc.org/pdf/Security_Guidance_for_Wastewater_Stormwater_Facilities.pdf

EPA, 2005. Emergency Response Tabletop Exercises http://www.epa.gov/safewater/watersecurity/tools/traininged/

EPA, 2004. National Incident Management System (NIMS) Training http://cfpub.epa.gov/safewater/watersecurity/home.cfm?program_id=8#nims

CHAPTER 3.0

# LOOKING FORWARD: ONGOING SECURITY ENHANCEMENT

## 3.1 Introduction

The first part of this report outlined a series of practical and effective security enhancements applicable within the operating conditions and resource constraints typical of small, rural, and/or tribal wastewater utilities. While these practices are not derived from top-down or consensus-based security standards, they are based on a broad review of standard practice and represent the collective wisdom of municipal leaders and small system operators with good, long-term security records and/or experience in the successful management of emergency situations. These practices are also consistent with the features of an "active and effective" security program identified by the National Drinking Water Advisory Council. (NDWAC, 2005) While it cannot be argued that implementation of these practices will result in a "best in class" security system, it is clear that implementation of these practices will improve the security of many small wastewater systems.

In addition to addressing immediate needs and opportunities, it is important that wastewater utilities and their host municipalities lay out an ongoing strategy for systematic enhancement and continual improvement of their security status. This chapter summarizes the steps that a utility or municipality should undertake to implement a sustainable program of ongoing security enhancement. As a first step, it is recommended that utilities and their parent municipality conduct a security vulnerability assessment or other formalized approach to the identification and characterization of community-level risks. Section 3.2, below, provides a summary of widely utilized approaches to vulnerability assessment. It is further recommended that utilities and/or municipalities conduct their vulnerability assessments to either draw upon or provide input to other important management activities, such as strategic planning and asset management. To facilitate such a cross-walk, Section 3.3 provides background on utility management practices that share steps with security planning. Further, it is recommended that utilities take steps to identify and describe important relationships with other municipal or regional entities that may influence or be influenced by wastewater security-related activities or emergency events. Section 3.4 includes a matrix to help utilities identify productive security-related interactions with other municipal services or infrastructure areas. Section 3.5 includes information on funding sources available to support utilities or municipalities interested in pursuing security-related improvements, programs, or related investments.

## 3.2 Vulnerability Assessment

The key to a long-term security strategy is to conduct a system-wide or municipal vulnerability assessment. The U.S. Environmental Protection Agency, leading wastewater associations, and other technical/professional bodies have developed guidance for the conduct of vulnerability assessments. In all cases, vulnerability assessments involve a listing of assets, assessment of potential risks, and careful consideration of how likely risks might impact key system components. Outlined below, vulnerability assessment approaches range from complex, multi-attribute, probability-based approaches, to relatively simple, checklist-based tools:

### 3.2.1 VSAT Methodology

The VSAT methodology was developed to facilitate identification of security-related risks to critical assets of wastewater systems, and to perform comparative analyses for alternative mitigation approaches. To characterize risk, a system's critical assets are paired with perceived security threats, and the likelihood and consequences for each pair are assessed. Threat-asset combinations with relatively higher probabilities of occurrence and/or more severe consequences correspond with greater security-related risks. Cost-benefit analyses performed within VSAT help utilities to focus available resources on the most significant risks.

### 3.2.2 RAM-W

The RAM-W methodology begins with the identification and ranking of critical system components using a process termed "pair wise comparison." The process helps utilities to identify system components that are critical for continued operation at suitable levels and helps the utility to prioritize security upgrades and/or modify policies and procedures to mitigate identified risks. RAM-W provides utilities with a way to allocate resources according to relative levels of risk.

### 3.2.3 Checklist Approaches

Several checklist-based approaches have been developed to assist utilities in assessing vulnerabilities. The *Asset Based Vulnerability Checklist for Wastewater Utilities*, produced by Association of Metropolitan Sewerage Agencies (AMSA, 2002), provides a means for systematically considering utility assets to identify and characterize potential vulnerabilities. Similarly, the National Environmental Training Center for Small Communities (NETCSC, 2002) and the National Rural Water Association (NRWA, 2004) each developed workbook-type approaches.

### 3.2.4 Conducting a Vulnerability Assessment: The Basics

Based upon the technical, financial, and managerial capabilities of most small wastewater systems, it may be unreasonable to attempt the type of comprehensive, formulaic approaches called for under VSAT and RAM-W. By way of reference, none of the utilities interviewed for this study had utilized either VSAT or RAM-W. However, the survey and workshop inputs suggest that a simplified, checklist approach is feasible, especially if conducted in close coordination with the municipality as a whole. Of the utilities interviewed for this study that had conducted a vulnerability assessment, approximately one-half had used an abbreviated template or tool provided by their state rural water association.

While various vulnerability assessment approaches appear suited for use by small wastewater systems, the underlying logic is quite consistent across all templates, and consists of five basic steps:

**Build a team** It is important that your vulnerability assessment be informed by multiple perspectives. Ideally, a vulnerability assessment team should include the operator, board members, owners, and operating contractors.

**Evaluate your system** Take time to identify different types of customers and consider their divergent needs.

**Inventory critical assets** Make a list of your system assets, including at least the following categories: personnel, treatment, power, collection, offices, and communications. For each asset provide a description and include information on the type, number, location, age, and condition of assets.

**Conduct a threat assessment** "Threat Assessment" is a fancy term that means: think carefully and make a list of potential hazards and things that can go wrong in and around your system and community. For each hazard, do your best to consider its frequency of occurrence, probability of occurrence, and magnitude of impact.

> **Asset Management** is a planning process that ensures that utilities and/or communities get the most value from their assts and have the financial resources to rehabilitate and replace them when necessary. Asset management also includes developing a plan to reduce costs while increasing the efficiency and reliability of system assets.
>
> **Strategic Planning** is a management concept that helps utilities address and prepare for both anticipated and unanticipated problems. Strategic planning utilizes asset management to evaluate a system's current situation, and also evaluates a system's financial and managerial state-of-affairs. It requires utility leaders to make fundamental decisions regarding their system's purpose, structure, and functions.

**Combine the Asset Inventory and Threat Assessment** Once you have made a list of assets and potential hazards, combine the two by thinking about how each hazard type could affect each of your major assets or asset categories. This should allow you to zero-in on the most vulnerable assets and begin to consider appropriate corrective actions.

As described in Sections 3.3 and 3.4, security vulnerability assessments can be conducted to feed into other utility and/or municipal-level planning and management activities.

## 3.3 Meshing Security Enhancement with Good Management Practices

Municipalities and wastewater utilities are subject to a range of business- and management-oriented process requirements, such as strategic planning (U.S. EPA, 2003a), asset management (U.S. EPA, 2003b), environmental planning (U.S. EPA, 1994), financial reporting under the Governmental Accounting Standards Board's Statement #34, state insurance underwriting audits, and various bonding reviews.

Efforts to assess vulnerabilities and enhance wastewater system security can and should be coordinated with other utility and municipal business management activities. Combined with security planning, processes such as strategic planning and asset management can be used to help frame forward-looking efforts to identify threats, opportunities and needs, and potential pathways to achieve goals and longer term objectives. Table 3-1, below, maps similarities and commonalities among asset management, strategic planning, and security-related vulnerability assessment.

**Table 3-1. Similarities Among Utility Management Practices.**

| Vulnerability Assessment & Emergency Response Planning Processes: Key Steps | Asset Management Process: Key Steps | Strategic Planning Process: Key Steps |
|---|---|---|
| 1. Inventory critical components<br>2. Identify vulnerabilities<br>3. Identify actions to address vulnerabilities<br>4. Prioritize actions | 1. Inventory assets<br>2. Prioritize assets<br>3. Develop asset management plan<br>4. Implement plan<br>5. Review & revise plan | 1. Develop strategic roadmap<br>2. Define area(s) of service<br>3. Assess technical, managerial, & financial capabilities<br>4. Identify options for fulfilling area(s) of service<br>5. Analyze & assess options<br>6. Implement options through an action plan<br>7. Evaluate options |

Based on this commonality of purpose and focus, the researchers recommend that utilities view security vulnerability assessment as an aspect of other good management practices, and not as a new and separate mandate. Integration of security with other ongoing management practices will help to make it a "living" aspect (as opposed to a mere paper exercise) of the utility's day-in and day-out business practice. Furthermore, a utility's chance of receiving grant or other financial assistance will be enhanced to the degree that funding proposals can identify multiple benefits, such as environmental protection and security enhancement.

## 3.4 Addressing Security as a Municipal or Regional Initiative

It is important to bear in mind that provision of wastewater services is only one of many responsibilities or infrastructure services provided by a community. This can both enable and constrain a utility's efforts to enhance system security. Other community needs can influence which wastewater projects are pursued and when they might be completed. For example, in a small community, collection system expansion or rehabilitation is likely to be tied to the street repair schedule and associated budgeting.

It is also important for wastewater utilities to work with other local and regional infrastructure and public service providers to identify key interdependencies, and to understand how failures in one service area could cause a cascade or ripple effect leading to damage or service disruption in other sectors. On the other hand, counterpart agencies and departments are also a good sourcc of mutual assistance, especially if they utilize similar technical skills and equipment. Wastewater utility management also has an opportunity to work with other local government agencies to train their employees to be aware of system components and their basic function, and more specifically, to be aware of unauthorized entry into manholes, catch basins, or curb inlets. Due to the nature of their jobs, municipal staff such as police, refuse haulers, and road crews travel frequently over a wide range of the municipality, and are thus well positioned to observe and report suspicious behavior or other irregular aspects of system performance. It is therefore important that wastewater utility leaders interact closely with other regional and/or municipal service providers when considering security-related upgrades, enhancements, or policy initiatives.

Table 3-2 is constructed to help wastewater utility staff and leaders identify opportunities to coordinate with other municipal or regional entities that may have similar security-related needs. The table lists common municipal service or infrastructure categories across the top; and potentially shared or overlapping operational characteristics along the vertical axis. The table is filled in with illustrative information about how a wastewater utility might interact with other units of infrastructure. For example, the bold, italicized cell indicates that a wastewater utility and a drinking water utility have elements of infrastructure in close physical proximity, suggesting the need and/or opportunity to coordinate security-related upgrades. The bottom row of the table, "Potential Areas of Collaboration," describes the relationship between infrastructure/service categories and operational characteristics in terms of 1) whether there are factors that would support a mutual aid or assistance relationship, 2) whether there are factors that support joint security-related training and/or emergency response planning, and 3) whether there may be opportunities for joint capital planning that could include security-related enhancements. It is recommended that utilities use the basic structure provided by Table 3-2, but populate the matrix with categories and relationships relevant to their own unique circumstances;

in other words, do not use the exhibit as is, but rework the relationships to assure their fit and relevance to local conditions.

**Table 3-2. Community and/or Regional Level Infrastructure and Services With Potentially Interrelated Operational Characteristics.**

| Operational Characteristics | Infrastructure or Service Category | | | | | | | | | | |
|---|---|---|---|---|---|---|---|---|---|---|---|
| | **Drinking Water Systems** | **Energy Generation Facilities** | **Energy Transmission or Distribution Infrastructure** | **Transportation Infrastructure** | **Facilities (e.g., government buildings, prisons, military installations)** | **Educational Institutions** | **Stadiums, Parks, & Public Gathering Areas** | **Communication & Technological Infrastructure** | **Hospitals & Medical Clinics** | **Law Enforcement** | **Fire, Ambulance, and EMT Services** |
| Wastewater System Customer | Yes, but reasonable tolerance for service disruption | Yes, but reasonable tolerance for service disruption | Not applicable | Yes, but reasonable tolerance for service disruption | Probably, major customer with moderate-to-limited ability to tolerate service disruption | Yes, major customer with limited ability to tolerate service disruption | Yes, major customer with limited ability to tolerate service disruption | Yes | Yes, major customer with virtually no ability to tolerate service disruption | Yes | Yes |
| Wastewater System Service Provider | Yes | Wastewater systems cannot operate without electricity | Wastewater systems cannot operate without electricity | Wastewater system operation, maintenance, and repair are hampered when standard transportation routes are impassable | No | No | No | Telecommunications systems are key to both routine and emergency operation of wastewater systems | No | Yes | Yes |
| Functionally Dependant Infrastructure | Potentially | Wastewater systems cannot operate without electricity | Wastewater systems cannot operate without electricity | Possibly, wastewater collection systems may utilize roadway bridges to cross water bodies | No | No | No | Telecommunications systems are key to both routine and emergency operation of wastewater systems | No | Yes | Yes |
| Spatially Overlapping Equipment or Infrastructure | ***DRINKING WATER DISTRIBUTION SYSTEMS AND WASTEWATER COLLECTION SYSTEMS ARE LIKELY TO BE SPATIALLY CONTIGUOUS*** | Sewage back-ups could impact electrical switching or generation equipment | Wastewater collection systems may overlap buried power cables. | Wastewater collection systems are often contiguous with or buried below transportation routes | Wastewater collection pipes and appurtenances may be located within or near government facilities, creating potential security access risk | No | Public easements (trails, bike paths) are often located along wastewater collection system right-of-ways | Communications infrastructure is often contiguous with wastewater collection pipes | No | Yes | Yes |
| Shared Political Leadership | Often operate under consolidated authority, subject to local governmental bodies (mayor, council) | Unlikely, municipally owned and operated electric utilities are not common | Not likely, electricity and natural gas pipelines and distribution facilities are usually privately owned and operated | Frequently operate as "sister" departments under a single municipal authority | Municipal, state, and/or federal government | Elementary and secondary schools, and community colleges are often operated by municipalities; colleges and universities are usually private or operated by State governmental bodies | Municipal government | Communications infrastructure is most often privately owned and operated | No | Municipal, state, and/or federal government | Municipal government |
| Utilization of Similar Equipment and/or Technical Expertise | Both utilize heavy equipment and require access to staff with mechanical, electrical, chemical, and computer-related skills | Both utilize heavy equipment and require access to staff with mechanical, electrical, and computer skills | Both utilize heavy equipment and require access to staff with mechanical, electrical, and computer skills | Both utilize heavy equipment and require access to staff with mechanical and computer skills | Both rely on similar technical expertise for perimeter protection and security. | Universities may provide technical and/or laboratory skills and capabilities; may have heavy equipment and skilled operators and mechanical staff | Similar utilization of municipal services, e.g., snow removal (where applicable), landscaping, street cleaning, etc. | Potential source of computer skills; experience and capability in customer notification | No | Yes | Yes |
| **Potential Areas of Collaboration** | Mutual aid and assistance agreement; provision of emergency supplies, staff, equipment; joint training and emergency response planning activities; partners in emergency procurement; potential collaboration under municipal-wide asset management inventory | Mutual aid and assistance agreement; staff; equipment | Mutual aid and assistance agreement; staff; equipment | Mutual aid and assistance agreement; provision of emergency staff and equipment; joint emergency procurement arrangements; potential collaboration under county or municipal asset management inventory; potential partners in citizen watch program | Mutual aid and assistance agreement; staff; equipment | Potential mutual aid and assistance partner; likely emergency response planning partner or stakeholder | Possible mutual aid and assistance partner; public areas may serve as gathering areas or shelters during emergency events; potential emergency response planning partner/stakeholder | Mutual aid and assistance agreement; staff; equipment; Government Emergency Telecommunications Service (GETS Program) | Staff; ICS | Key role in municipal incident command system (ICS); partner in citizen watch-type program | ICS; Emergency Communication |

## 3.5 Identifying and Obtaining Funding for Security Enhancements

While the security practice areas identified in Chapter 2 were selected partially because they can be implemented inexpensively, long-term programs of security enhancement identified through vulnerability assessments and community planning may be expensive. It is likely that small communities and wastewater systems will require financial assistance to implement long-term security programs. Many sources provide grants and loans to utilities and municipalities to assist in security and emergency response needs. In some cases, funding must be sought through State primacy agencies or other state agencies.

Table 3-3 provides sources of grants and loans available to utilities or municipalities planning security-related investments.

**Table 3-3. Federal Funding Sources that Wastewater Utilities and Municipalities Can Access to Obtain Funding to Support Security-Related Projects and Programs.**

| Title | URL | Description |
|---|---|---|
| Hardship Grants Program for Rural Communities | www.epa.gov/owm | Many rural communities cannot afford the full cost of State Revolving Fund loans. Communities such as these can seek assistance through U.S. EPA's Hardship Grants program, which helps small, disadvantaged rural communities with fewer than 3,000 people address wastewater treatment needs. |
| Department of Agriculture, Rural Utilities Service (RUS) | www.usda.gov/rus/water/ | The Rural Utilities Service provides grants and loans to rural communities with fewer than 10,000 people for wastewater and other utility projects. The RUS funds may be used to install, repair, improve, or expand rural wastewater or drinking water facilities. |
| U.S. Department of Health and Human Services, Indian Health Services (IHS) | http://www.dsfc.ihs.gov/mission.cfm#4 | The IHS Sanitation Facilities Construction (SFC) program provides funding to construct sanitation and wastewater disposal facilities. |
| U.S. Environmental Protection Agency (EPA), State Revolving Fund (SRF) | www.epa.gov/owm/cwfinance/cwsrf/contacts.htm | Clean Water State revolving find programs provide about $4 billion annually to fund water quality protection projects for wastewater treatment. To receive funding, a project must be in the host state's intended use plan. |
| U.S. Federal Emergency Management Agency (FEMA), Pre-Disaster Mitigation Program (PDM) | www.fema.gov/fima/pdm.shtm | The PDM program provides funds to states, territories, Indian tribal governments, communities, and universities for hazard mitigation planning and implementation of mitigation projects prior to a disaster event. |
| U.S. Federal Emergency Management Agency (FEMA), Hazard Mitigation Grant Program (HMGP) | www.fema.gov/fima.hmgp | The HMGP assists states and local communities in implementing long-term hazard mitigation measures following a major disaster declaration. |

CHAPTER 4.0

# SECURITY IN THE CONTEXT OF A HYPOTHETICAL COMMUNITY

## 4.1 Introduction

Ruraltown, population 1,500, is located in a low-lying coastal region in the northwestern United States.[1] The region is subject to a variety of extreme weather events, wildfires, and occasional seismic activity. A neighboring community was destroyed by a tsunami in the early 1960s. About half of the families in the community own and run farms. Another 100 people work in a mine in Quarryville, 15 miles away. Employers in Ruraltown include a seafood processing plant, a saw mill, several small stores and restaurants, and a movie theater. The nearby state park provides summer jobs. The average household income is about $25,000 a year. The average home is worth approximately $70,000.

The nearest town, Hoppington, is two miles away and has 2,500 residents. The nearest city, Metro Center, is 25 miles away and has a population of 300,000 people. Ruraltown, Hoppington, and Metro Center are all part of Carroll County.

Ruraltown has no full-time mayor, no town attorney, and no environmental planner. Ruraltown does have two full-time paid government employees. One is the town clerk, who keeps track of state and federal requirements (and does the accompanying paperwork), collects town taxes, and is part of the volunteer fire department. The town also employs a full-time public works director to operate the drinking water and wastewater treatment plants and coordinate the municipal solid waste program. In addition to the two full-time employees, the town employees a part-time wastewater system operator, but the individual has not yet been certified by the state.

Decisions in Ruraltown are made at a twice-yearly town meeting. The town meeting is run by the part-time mayor, who has served as Ruraltown's mayor for more than 30 years. Ruraltown also has a part-time town council. The town council presents information at the town meeting and gives advice on what the town should do. The council also meets every month to take care of business that needs attending to between town meetings.

About 200 residents get their drinking water from private wells instead of the public water supply. Some private wells have been contaminated by nearby septic tank systems; others could be contaminated in the future by sources such as underground storage tanks, improperly disposed of used oil and household hazardous waste, runoff from farms, and contaminants leached from the landfill.

---

[1] *The example of Ruraltown is adapted from EPA guidance dealing with environmental planning on a community or regional scale, see EPA, 1994.*

Ruraltown has problems in the area of wastewater treatment. Approximately three-quarters of the homes and businesses in the town have sewer hookups, and the wastewater treatment facility is old and sometimes violates permit levels. The rest of the homes and a few businesses use septic systems. The Ruraltown utility operates on an annual budget of approximately $180,000. About half of the budget is dedicated to salary and electric utilities; little is left for "discretionary" expenditures. The utility does not collect enough revenue to acquire an appreciable operational reserve.

Despite significant constraints and limitations, the Ruraltown Public Works Director worked through her mayor and council to launch a cooperative planning effort with neighboring entities. As part of this effort, the planning team conducted a simple security-related vulnerability assessment, including designation of likely security threats and "at risk" infrastructure components. Some key aspects of the planning exercise are outlined below:

## 4.2 Joint Capital Planning and Infrastructure Development

Economic development in Metro Center is beginning to spill into Hoppington and Ruraltown, which are becoming "bedroom" suburbs of Metro Center. Possibilities exist for economic development initiatives that may support infrastructure enhancement. Coordination opportunities may occur during roadway construction, drinking water system expansion, or other municipal ventures. Interconnections between wastewater systems may be a feasible outcome of development activities. Other jurisdictions may find interconnects an attractive way to increase the scale (and hence reduce unit costs) as they pursue expansion or upgrade initiatives.

On another front, Ruraltown and Hoppington have developed a joint proposal for grading and installation of risers to help reduce the incidence of I&I events for six of their eight lift stations. In preparation for this project, the two communities plan to commission the services of a flood plain expert to review the siting and design of sewage lift stations and to plan by-pass approaches for vulnerable sites. This project was submitted and accepted under the state's SRF intended use plan. Ruraltown is also in the process of completing a proposal for RUS funding to obtain wireless SCADA systems for two of its older and more remote lift stations.

## 4.3 Mutual Aid and Assistance and Provision of Emergency Surge Capacity

Ruraltown is part of a county-wide mutual aid and assistance program that includes Carroll County law enforcement, fire rescue, and emergency management agencies; Carroll County and Metro City health and public works departments; and water, solid waste, electric, and gas utilities in Hoppington and Metro City. Ruraltown and Hoppington have a handshake agreement to share system operators during periods of emergency need. Ruraltown is considering joining the statewide WARN program.

## 4.4 Emergency Response Planning and Incident Command Systems (ICS)

Utilities should be proactive in working with jurisdictional partners (e.g., mayors, public utility directors) to determine ICS roles and expectations with respect to their roles in county, township, or municipal ERPs. This is especially crucial in this case due to the part-time status of Ruraltown's political leadership.

## 4.5 Emergency Procurement

The part-time nature of Ruraltown's political leadership makes it likely that decision-making could become problematic during an emergency event. To undertake rapid and effective response to and recovery from disasters, the town has adopted a resolution that allows utility staff to procure supplies and services quickly and outside the business-as-usual processes. This procedure addresses emergency purchases that may be necessary under extreme conditions where high-level approvals are not achievable and where communication networks are out of service, and includes a provision for "after-the-fact" approval/authorization.

## 4.6 Training

Ruraltown wastewater agency staff conduct training activities, including quarterly breakfast meetings and "tailgate exercises," in cooperation with colleagues in Hoppington, Metro City, and Carroll County. Ruraltown's mayor and council members receive an annual facility tour and briefing on key start-up and shut-down procedures. Also on an annual basis, utility staff participate in a county-wide ERP activation exercise.

# APPENDIX

# SMALL WASTEWATER SYSTEM SECURITY BEST PRACTICES SURVEY

Thank you for participating in this confidential survey. It is part of a Water Environment Research Foundation (WERF) project to develop a security enhancement and emergency preparedness tool kit for small wastewater system operators. Based partially on a review of "best practices" from systems such as yours, the tool kit will assist utilities in implementing viable security improvements, emergency response activities, risk communication plans, and coordination efforts with local communities and other utility systems. The tool kit will be made available through WERF.

Please review the survey questions in advance of your scheduled telephone interview. Note that space is provided after each question for you to note key points or other important information. You are encouraged to seek input from colleagues.

1. Has your utility experienced any natural or human-caused disaster during the past 10 years that resulted in system upset or other non-routine response? ☐ **YES** ☐ **NO**

***NOTES:***

2a. Does your utility have a written emergency response plan? ☐ **YES** ☐ **NO**

2b. If "yes," does this plan address specific types of emergencies? ☐ **YES** ☐ **NO**

*NOTES:*

3a. Has your utility performed a vulnerability assessment? ☐ **YES** ☐ **NO** (skip to question 5)

3b. If "yes," did you utilize any guidance materials or tools to conduct and compile your vulnerability assessment? ☐ **YES** ☐ **NO**

*NOTES:*

4. *(Skip this question, if your answer to Question #3 was "no.")* In developing your vulnerability assessment, did you consider particular types of threats and their likely impact on specific system assets? ☐ **YES** ☐ **NO**

*NOTES:*

5. Has your organization's vulnerability assessment been updated within the past two years? ☐ **YES** ☐ **NO**

*NOTES:*

6. Does your utility/municipality have an emergency response plan for system recovery? ☐ **YES** ☐ **NO**

*NOTES:*

7. Has your system's emergency response and recovery plan ever been activated since it was developed? ☐ **YES** ☐ **NO**

*NOTES:*

8a. Are you familiar with the concept of an Incident Command System (ICS)? ☐ **YES** ☐ **NO**

8b. *(Skip this question if your answer to Question #8a was "no.")* Is the organizational structure of your emergency response and recovery plan consistent with the ICS used by other responders (e.g., fire, police) in your area? ☐ **YES** ☐ **NO**

*NOTES:*

| | | |
|---|---|---|
| 9a. Is your utility/municipality involved in any inter-organizational cooperative mechanisms between water agencies/municipalities, such as mutual aid and assistance? | ☐ **YES** | ☐ **NO** (skip to question 10) |
| 9b. If "yes," are there aspects of this arrangement that could be improved? | ☐ **YES** | ☐ **NO** |

*NOTES:*

| | | |
|---|---|---|
| 9c. Are there aspects of this arrangement that work very well? | ☐ **YES** | ☐ **NO** |

*NOTES:*

10. How do you rate your utility's emergency response preparedness? (Circle one)

**Excellent**

**Good but could be enhanced**

**Fair, but needs a number of improvements**

**Improvement needed, needs a number of enhancements**

**Not adequate**

*NOTES:*

11. Do employees of your utility have designated emergency roles or responsibilities? ☐ **YES** ☐ **NO**

*NOTES:*

12. Have the emergency personnel in your utility been given training on your organization's emergency response and recovery plan? ☐ **YES** ☐ **NO**

*NOTES:*

13. Does your utility conduct periodic or regular emergency drills or other exercises? ☐ **YES** ☐ **NO**

*NOTES:*

14. Do you practice emergency response activation with outside agencies (police, fire, public health)? ☐ **YES** ☐ **NO**

*NOTES:*

15a. Has your utility undertaken a formal "asset management" process in which all system assets were inventoried? ☐ **YES** ☐ **NO**

*NOTES:*

15b. Has your utility undertaken a formal "asset management" process in which all system assets were prioritized? ☐ **YES** ☐ **NO**

*NOTES:*

For the following, please consider whether your utility's current status seems mostly adequate or inadequate:

1. Protection of treatment plant assets? ☐ **Mostly adequate** ☐ **Inadequate**

2. Protection of collection system assets? ☐ **Mostly adequate** ☐ **Inadequate**

3. Perimeter control? ☐ **Mostly adequate** ☐ **Inadequate**

4. Hazardous material control? ☐ **Mostly adequate** ☐ **Inadequate**

5. Arrangements with suppliers? ☐ **Mostly adequate** ☐ **Inadequate**

6. Addressing potential security-related issues regarding system personnel? ☐ **Mostly adequate** ☐ **Inadequate**

7. Protection of computers and computerized operations? ☐ **Mostly adequate** ☐ **Inadequate**

8. Protection of maps, records, and information? ☐ **Mostly adequate** ☐ **Inadequate**

*NOTES:*

Of the topics listed below, are there security-related practices that you feel are especially effective and worthy of broad emulation within the community of small wastewater utilities? Please describe.

1. Protection of treatment assets?

2. Protection of collection system assets?

3. Perimeter control?

4. Hazardous material control?

5. Arrangements with suppliers?

6. Addressing potential security-related issues regarding system personnel?

7. Protection of computers and computerized operations?

8. Protection of maps, records, and information?

# References

ALA, 2004. *Wastewater System Performance Assessment Guideline*. American Lifelines Alliance. http://www.americanlifelinesalliance.org.

Anderson, Robert M. and Shultz, C.R. 2005. *Legal Liability of Water Systems for Terrorist Induced Harm.* Washington, D.C.: LeBoeuf, Lamb, Greene & MacRae, L.L.P.

ASCE, 2004. *Voluntary Security Guidance for Wastewater/Stormwater Utilities*. Reston, VA: American Society of Civil Engineers.

ASCE, 2006. *Guidelines for the Physical Security of Wastewater/Stormwater Utilities*. Reston, VA: American Society of Civil Engineers. (Published in association with the American Water Works Association and the Water Environment Federation)

ASIS, 2003. *Suggested Anti-Terrorism Security Measures*. Alexandria, VA: ASIS International. http://www.asisonline/antiterror.html.

ASIS, 2007. *Major Planning Considerations Checklist*. Alexandria, VA: ASIS International. http://www.asisonline.org.

AMSA, 2002. *Asset Based Vulnerability Checklist for Wastewater Utilities*, Washington, D.C.: Association of Metropolitan Sewerage Agencies.

Bravata, D.M., McDonald, K.M., Owens, D.K., Wilhelm, E.R., Brandeau, M.L., Zaric, G.S., Holty, J.E.C., Liu, H., and Sandaram, V., 2004. *Regionalization of Bioterrorism Preparedness and Response*. Summary, Evidence Report/Technology Assessment: Number 46. AHRQ Publication Number 04-E016-1, April 2004. Rockville, MD: Agency for Healthcare Research and Quality. http://www.ahrq.gov/clinic/epcsums/bioregsum.htm.

FEMA, 2000. *Emergency Response to Terrorism Job Aid.* Washington, D.C.: Federal Emergency Management Agency, United States Fire Administration, National Fire Academy, and United States Department of Justice.

FlaWARN. 2006. *Emergency Response and Preparedness Best Practices for Florida's Water and Wastewater Systems*. Gainesville, FL: University of Florida Center for training, Research and Education for Environmental Occupations.

GASB, 2005. Governmental Accounting Standards Board Statement No. 34, Basic Financial Statements – and Management's Discussion and Analysis – for Sate and Local Governments. Washington, D.C.: Governmental Accounting Standards Board.

GAO, 2005. *Wastewater Facilities: Experts Views on How Federal Funds Should be Spent to Improve Security*. Washington, D.C.: United States Government Accountability Office.

Gasteyer, Stephen. 2005. *Fredonia and Columbus City Achieve Wastewater Treatment through Regionalization*. Washington, D.C.: Rural Community Assistance Partnership.

IDNR, 1996. *Self-Assessment Manual for Iowa Water System Viability: Rural Water Association-Owned and Municipality-Owned Systems*. Iowa Department of Natural Resources.

MIIC, 2007. *Infrastructure Status Report: Massachusetts Wastewater Facilities*. Boston, MA: Massachusetts Infrastructure Investment Coalition.

Moyer, Jack. 2007. *10 Steps to Preparedness: No Wastewater Utility's Emergency Response Plan is Complete Unless it Addresses these 10 Steps*. Water, Environment, and Technology, March 2007, pages 72-76.

NCPC, 2002. *Starting a Neighborhood Watch*. Washington, D.C.: National Crime Prevention Council. Accessed at http://www.ncpc.org/cms/cms-upload/ncpc/files/nwstart.pdf.

NCSC, 2008. *Community Leadership in a Risky World: Risk Management Starter Kit*. Washington, D.C.: National Center for Small Communities. Accessed at http://www.smallcommunities.org/.

NDWAC, 2005. *Recommendations of the National Drinking Water Advisory Council to the U.S. Environmental Protection Agency on Water Security Practices, Incentives, and Measures*. Washington, D.C.: National Drinking Water Advisory Council.

NESCSC, 2002. *Protecting Your Community's Assets: A Guide for Small Wastewater Systems*. Morgantown, WV: National Environmental Services Center for Small Communities.

NLC, 2002. *Homeland Security: Practical Tools for Local Governments*. Washington, D.C.: National League of Cities.

NRC, 1997. *Safe Water From Every Tap*. Washington, D.C.: National Research Council

NRWA, 2002. *Rural and Small Water and Wastewater System Emergency Response Plan Template*. Duncan, Oklahoma: National Rural Water Association.

NRWA, 2002. *Expert Panel on Anti-Terrorism for Small Rural Water Systems*. Washington, D.C.: National Rural Water Association.

NRWA, 2004. *Security Vulnerability Self-Assessment Guide for Small Wastewater Systems*. Duncan, Oklahoma: National Rural Water Association.

PERI, 2002. *Characteristics of Effective Emergency Management Organizational Structures*. Fairfax, Virginia: Public Entity Risk Institute.

Regnier, John E. 1987. *Survey Characterizes the Small Water System*. Water Engineering & Management, October 1987, Pages 38-40.

USDA, 2007. *Loan and Grant Approval Conditions for [specified] System Approval Project*. RUS Bulletin 1780-19. Washington, D.C.: United States department of Agriculture, Rural

Utilities Service.

U.S. EPA, 1994. *Environmental Planning for Small Communities: A Guide for Local Decision-Makers*. Washington, D.C.: U.S. Environmental Protection Agency, Office of Research and Development. Publication Number: EPA/625/R-94/009.

U.S. EPA, 2001. *What Wastewater Utilities Can Do Now to Guard Against Terrorist and Security Threats.* Washington, D.C.: U.S. Environmental Protection Agency, Office of Wastewater Management.

U.S. EPA. 2001. *Use of the Drinking Water State Revolving Fund (DWSRF) to implement Security Measures at Public Water Systems*. Washington, D.C.: U.S. Environmental Protection Agency, Office of Water. Publication Number: EPA816-F-02-040.

U.S. EPA, 2003a. *Strategic Planning: A Handbook for Small Water Systems: One of the Simple Tools for Effective Performance (STEP) Guide Series*. Washington, D.C.: U.S. Environmental Protection Agency, Office of Water.

U.S. EPA, 2003b. *Asset Management: A Handbook for Small Water Systems: One of the Simple Tools for Effective Performance (STEP) Guide Series*. Washington, D.C.: U.S. Environmental Protection Agency, Office of Water.

U.S. EPA, 2003. *Water Watchers: Helping to Protect Your Local Water System*. Washington, D.C.: U.S. Environmental Protection Agency, Office of Water. Publication Number: EPA-810-F-03-006.

U.S. EPA, 2003. *Fact Sheet: Use of the Clean Water State Revolving Fund to Implement Security Measures at Publicly-Owned Wastewater Treatment Works*. Washington, D.C.: U.S. Environmental Protection Agency, Office of Water. Publication Number: EPA 832-F-03-002.

U.S. EPA, 2005. *Drinking Water Security for Small Systems Serving 3,300 or Fewer Persons: One of the Simple Tools for Effective Performance (STEP) Guide Series*. Washington, D.C.: U.S. Environmental Protection Agency, Office of Water. Publication Number: EPA 817-R-05-001.

U.S. EPA, 2005. *Security Information Collaboratives: A Guide for Water Utilities*. Cincinnati, OH: U.S. Environmental Protection Agency, Office of Research and Development, National Homeland Security Research Center. Publication Number: EPA/625/R-05/002.

U.S. EPA, 2005. *Emergency Response Tabletop Exercises for Drinking Water and Wastewater Systems*. Washington, D.C.: U.S. Environmental Protection Agency, Office of Water, Water Security Division. Public Number: EPA 817-C-05-001.

U.S. EPA, 2007. *Effective Risk and Crisis Communication during Water Security Emergencies: Summary Report of EPA Sponsored Message Mapping Workshops*. Cincinnati, OH: U.S. Environmental Protection Agency, Office of Research and development, National Homeland Security Research Center. Publication Number: EPA/600/R-07/027.

U.S. EPA, 2007. *EPA Federal Funding Sources for Small Community Wastewater Systems*. Washington, D.C.: U.S. Environmental Protection Agency, Office of Wastewater Management, Municipal Support Division. Publication Number: EPA 832-F-97-004.

WDHFS, 2004. *Emergency Preparedness in Wisconsin: Results of Five Focus Groups.* Madison, Wisconsin: Wisconsin Department of Health and Family Services, Division of Public Health.

WEF, 2001. *Privatization: Public or Private Management and Operation of Wastewater Treatment Utilities*. Alexandria, VA: Water Environment Research Foundation. Accessed on 4/5/07 from http://www.wef.org/GovernmentAffairs/PolicyPositionStatements/.

WERF, 2004. *Emergency Response Plan Guidance for Wastewater Systems*. Alexandria, VA: Water Environment Research Foundation.

## WASTEWATER UTILITY

### Alabama
Montgomery Water Works & Sanitary Sewer Board

### Alaska
Anchorage Water & Wastewater Utility

### Arizona
Avondale, City of
Glendale, City of, Utilities Department
Mesa, City of
Peoria, City of
Phoenix Water Services Dept.
Pima County Wastewater Management
Safford, City of
Tempe, City of

### Arkansas
Little Rock Wastewater Utility

### California
Central Contra Costa Sanitary District
Corona, City of
Crestline Sanitation District
Delta Diablo Sanitation District
Dublin San Ramon Services District
East Bay Dischargers Authority
East Bay Municipal Utility District
El Dorado Irrigation District
Fairfield-Suisun Sewer District
Fresno Department of Public Utilities
Inland Empire Utilities Agency
Irvine Ranch Water District
Las Gallinas Valley Sanitary District
Las Virgenes Municipal Water District
Livermore, City of
Los Angeles, City of
Los Angeles County, Sanitation Districts of
Napa Sanitation District
Novato Sanitary District
Orange County Sanitation District
Palo Alto, City of
Riverside, City of
Sacramento Regional County Sanitation District
San Diego Metropolitan Wastewater Department, City of
San Francisco, City & County of
San Jose, City of
Santa Barbara, City of
Santa Cruz, City of
Santa Rosa, City of
South Bayside System Authority
South Coast Water District
South Orange County Wastewater Authority
South Tahoe Public Utility District
Stege Sanitary District
Sunnyvale, City of
Union Sanitary District
West Valley Sanitation District

### Colorado
Aurora, City of
Boulder, City of
Greeley, City of
Littleton/Englewood Water Pollution Control Plant
Metro Wastewater Reclamation District, Denver

### Connecticut
Greater New Haven WPCA
Stamford, City of

### District of Columbia
District of Columbia Water & Sewer Authority

### Florida
Broward, County of
Fort Lauderdale, City of
Jacksonville Electric Authority (JEA)
Miami-Dade Water & Sewer Authority
Orange County Utilities Department
Pinellas, County of
Reedy Creek Improvement District
Seminole County Environmental Services
St. Petersburg, City of
Tallahassee, City of
Toho Water Authority
West Palm Beach, City of

### Georgia
Atlanta Department of Watershed Management
Augusta, City of
Clayton County Water Authority
Cobb County Water System
Columbus Water Works
Fulton County
Gwinnett County Department of Public Utilities
Savannah, City of

### Hawaii
Honolulu, City & County of

### Idaho
Boise, City of

### Illinois
Decatur, Sanitary District of
Greater Peoria Sanitary District
Kankakee River Metropolitan Agency
Metropolitan Water Reclamation District of Greater Chicago
Wheaton Sanitary District

### Indiana
Jeffersonville, City of

### Iowa
Ames, City of
Cedar Rapids Wastewater Facility
Des Moines, City of
Iowa City

### Kansas
Johnson County Wastewater
Unified Government of Wyandotte County/ Kansas City, City of

### Kentucky
Louisville & Jefferson County Metropolitan Sewer District
Sanitation District No. 1

### Louisiana
Sewerage & Water Board of New Orleans

### Maine
Bangor, City of
Portland Water District

### Maryland
Anne Arundel County Bureau of Utility Operations
Howard County Bureau of Utilities
Washington Suburban Sanitary Commission

### Massachusetts
Boston Water & Sewer Commission
Massachusetts Water Resources Authority (MWRA)
Upper Blackstone Water Pollution Abatement District

### Michigan
Ann Arbor, City of
Detroit, City of
Holland Board of Public Works
Saginaw, City of
Wayne County Department of Environment
Wyoming, City of

### Minnesota
Rochester, City of
Western Lake Superior Sanitary District

### Missouri
Independence, City of
Kansas City Missouri Water Services Department
Little Blue Valley Sewer District
Metropolitan St. Louis Sewer District

### Nebraska
Lincoln Wastewater & Solid Waste System

### Nevada
Henderson, City of
Las Vegas, City of
Reno, City of

### New Jersey
Bergen County Utilities Authority
Ocean County Utilities Authority

### New York
New York City Department of Environmental Protection

### North Carolina
Charlotte/Mecklenburg Utilities
Durham, City of
Metropolitan Sewerage District of Buncombe County
Orange Water & Sewer Authority
University of North Carolina, Chapel Hill

### Ohio
Akron, City of
Butler County Department of Environmental Services
Columbus, City of
Metropolitan Sewer District of Greater Cincinnati
Montgomery, County of
Northeast Ohio Regional Sewer District
Summit, County of

### Oklahoma
Oklahoma City Water & Wastewater Utility Department
Tulsa, City of

### Oregon
Albany, City of
Clean Water Services
Eugene, City of
Gresham, City of
Portland, City of Bureau of Environmental Services
Lake Oswego, City of
Oak Lodge Sanitary District
Water Environment Services

### Pennsylvania
Hemlock Municipal Sewer Cooperative (HMSC)
Philadelphia, City of
University Area Joint Authority

### South Carolina
Charleston Water System
Mount Pleasant Waterworks & Sewer Commission
Spartanburg Water

### Tennessee
Cleveland Utilities
Murfreesboro Water & Sewer Department
Nashville Metro Water Services

### Texas
Austin, City of
Dallas Water Utilities
Denton, City of
El Paso Water Utilities

Fort Worth, City of
Houston, City of
San Antonio Water System
Trinity River Authority

**Utah**

Salt Lake City Corporation

**Virginia**

Alexandria Sanitation Authority
Arlington, County of
Fairfax County
Hampton Roads Sanitation District
Hanover, County of
Henrico, County of
Hopewell Regional Wastewater Treatment Facility
Loudoun Water
Lynchburg Regional Wastewater Treatment Plant
Prince William County Service Authority
Richmond, City of
Rivanna Water & Sewer Authority

**Washington**

Everett, City of
King County Department of Natural Resources
Seattle Public Utilities
Sunnyside, Port of
Yakima, City of

**Wisconsin**

Green Bay Metro Sewerage District
Kenosha Water Utility
Madison Metropolitan Sewerage District
Milwaukee Metropolitan Sewerage District
Racine, City of
Sheboygan Regional Wastewater Treatment
Wausau Water Works

**Water Services Association of Australia**

ACTEW Corporation
Barwon Water
Central Highlands Water
City West Water
Coliban Water Corporation
Cradle Mountain Water
Gippsland Water
Gladstone Area Water Board
Gold Coast Water
Gosford City Council
Hunter Water Corporation
Logan Water
Melbourne Water
Onstream
Power & Water Corporation
SEQ Water
South Australia Water Corporation
South East Water Limited
Sunshine Coast Water
Sydney Catchment Authority
Sydney Water
Wannon Regional Water Corporation
Watercare Services Limited (NZ)
Water Corporation
Water Distribution Brisbane City Council
Western Water
Yarra Valley Water

**Canada**

Edmonton, City of/Edmonton Waste Management Centre of Excellence
Lethbridge, City of
Regina, City of, Saskatchewan
Toronto, City of, Ontario
Winnipeg, City of, Manitoba

## STORMWATER UTILITY

**California**

Fresno Metropolitan Flood Control District
Los Angeles, City of, Department of Public Works
Monterey, City of
San Francisco, City & County of
Santa Rosa, City of
Sunnyvale, City of

**Colorado**

Aurora, City of
Boulder, City of

**Florida**

Orlando, City of

**Iowa**

Cedar Rapids Wastewater Facility
Des Moines, City of

**Kansas**

Lenexa, City of
Overland Park, City of

**Kentucky**

Louisville & Jefferson County Metropolitan Sewer District

**Maine**

Portland Water District

**North Carolina**

Charlotte, City of, Stormwater Services

**Pennsylvania**

Philadelphia, City of

**Tennessee**

Chattanooga Stormwater Management

**Texas**

Harris County Flood Control District, Texas

**Washington**

Bellevue Utilities Department
Seattle Public Utilities

## STATE

Connecticut Department of Environmental Protection
Kansas Department of Health & Environment
New England Interstate Water Pollution Control Commission (NEIWPCC)
Ohio Environmental Protection Agency
Ohio River Valley Sanitation Commission
Urban Drainage & Flood Control District, CO

## CORPORATE

ADS LLC
Advanced Data Mining International
AECOM
Alan Plummer & Associates
Alpine Technology Inc.
Aqua-Aerobic Systems Inc.
Aquateam–Norwegian Water Technology Centre A/S
ARCADIS
Associated Engineering
Bernardin Lochmueller & Associates
Black & Veatch
Blue Water Technologies, Inc.
Brown & Caldwell
Burgess & Niple, Ltd.
Burns & McDonnell
CABE Associates Inc.
The Cadmus Group
Camp Dresser & McKee Inc.
Carollo Engineers Inc.
Carpenter Environmental Associates Inc.
CET Engineering Services
CH2M HILL
CRA Infrastructure & Engineering
CONTECH Stormwater Solutions
D&B/Guarino Engineers, LLC
Damon S. Williams Associates, LLC
Ecovation
EMA Inc.
Environmental Operating Solutions, Inc.
Environ International Corporation
Fay, Spofford, & Thorndike Inc.
Freese & Nichols, Inc.
ftn Associates Inc.
Gannett Fleming Inc.
Garden & Associates, Ltd.
Geosyntec Consultants
GHD, Inc.
Greeley and Hansen LLC
Hazen & Sawyer, P.C.
HDR Engineering Inc.
HNTB Corporation
Hydromantis Inc.
HydroQual Inc.
Infilco Degremont Inc.
Jason Consultants LLC Inc.
Jordan, Jones, & Goulding Inc.
KCI Technologies Inc.
Kelly & Weaver, P.C.
Kennedy/Jenks Consultants
Larry Walker Associates
LimnoTech Inc.
Lombardo Associates, Inc.
The Low Impact Development Center Inc.
Malcolm Pirnie Inc.
Material Matters, Inc.
McKim & Creed
MWH
NTL Alaska, Inc.
O'Brien & Gere Engineers Inc.
Odor & Corrosion Technology Consultants Inc.
Parametrix Inc.
Parsons
Post, Buckley, Schuh & Jernigan
Praxair, Inc.
RMC Water & Environment
Ross & Associates Ltd.
SAIC
Siemens Water Technologies
The Soap & Detergent Association
Smith & Loveless, Inc.
Southeast Environmental Engineering, LLC
Stearns & Wheler, LLC
Stone Environmental Inc.
Stratus Consulting Inc.
Synagro Technologies Inc.
Tetra Tech Inc.
Trojan Technologies Inc.
Trussell Technologies, Inc.
URS Corporation
Wallingford Software
Westin Engineering Inc.
Wright Water Engineers
Zoeller Pump Company

## INDUSTRY

American Electric Power
American Water
Anglian Water Services, Ltd.
Chevron Energy Technology
The Coca-Cola Company
Dow Chemical Company
DuPont Company
Eastman Chemical Company
Eli Lilly & Company
InsinkErator
Johnson & Johnson
Merck & Company Inc.
Procter & Gamble Company
Suez Environment
United Utilities North West (UUNW)
United Water Services LLC
Veolia Water North America

**Note: List as of 11/10/09**

IWA Publishing's authorised EU representative for General Product Safety Regulations is Diane D'Arras, 15 rue Duret, 75116 Paris, France, e-mail: safety@iwap.co.uk.

Printed and bound by CPI Group (UK) Ltd, Croydon, CR0 4YY
12/05/2026
02108881-0007